A Guide for Statistical Tests and Interpretations with SPSS

A Guide for Statistical Tests and Interpretations with SPSS is designed for students taking basic and advanced courses in statistics, taking an integrative and practical approach to learning statistics. It guides students through navigating SPSS outputs and writing quantitatively, dealing with technical and substantive interpretations without resorting to complex mathematical formulae.

Starting from the basics of quantitative research methods and discussing descriptive and inferential statistical tests, this book provides a unique perspective of data analysis with SPSS. It makes a conscious effort to explore the various statistical methods one can use to dissect a data set using basic or advanced statistical techniques to achieve the best outcome. It covers the practical questions that arise while doing an assignment, final paper, or thesis – showing students how to proceed to the next step in their interpretation and analysis. It will provide quantitative methodology or data analysis students with core interpretations of SPSS outputs for key statistical tests. It will also demonstrate how to select and report the key trends and patterns of the data using descriptive and inferential statistics, the requirements and/or assumptions of each test, as well as the precise language to use for reporting on each test.

With SPSS screenshots and step-by-step advice, this book will be useful for all undergraduate and graduate students in the social sciences and humanities, as a supplemental textbook to provide practical guidance on moving through all steps of statistical testing and analysis.

Arshia U. Zaidi is an Associate Professor in the Faculty of Social Science & Humanities Criminology & Justice at Ontario Tech University (OTU), Canada, teaching statistics to both undergraduate and graduate students.

A Guide for Statistical Tests and Interpretations with SPSS

Arshia U. Zaidi

 Routledge
Taylor & Francis Group

NEW YORK AND LONDON

Designed cover image: Getty

First published 2025
by Routledge
605 Third Avenue, New York, NY 10158

and by Routledge
4 Park Square, Milton Park, Abingdon, Oxon OX14 4RN

Routledge is an imprint of the Taylor & Francis Group, an informa business

© 2025 Arshia U. Zaidi

Library of Congress Cataloging-in-Publication Data
A catalog record for this title has been requested

ISBN: 978-1-032-10520-8 (hbk)
ISBN: 978-1-032-10210-8 (pbk)
ISBN: 978-1-003-21569-1 (ebk)

DOI: 10.4324/9781003215691

Typeset in Optima
by Taylor & Francis Books

My first book as an academic must first and foremost be dedicated to my professors and colleagues, who mentored and supported my academic endeavors always and channeled my scholarly energies positively. Secondly, my folks, especially my father, Urooj A. Zaidi, who always told me a degree and books hold more value than any superfluous commodity in the world. Thirdly, my beating hearts: Fawad and Zoha who have been nothing but the fortitudinous bearers of my academic stressors and torch, with their inundated countless support, particularly on the long hard days of academic being. Finally, to my numerous undergraduate and graduate students whom I taught statistics to and discussed statistics with. This book is an 18-year compilation of all the statistical lectures, discussions, and conversations we had each semester. My students pushed me to do this for the adoration of statistical interpretations and numerical storytelling of data.

Contents

Figures

Boxes

Preface

Most individuals reading this book are mostly reading it because statistics is a core course and their preference would rather be doing something else, then "doing stats". In my years of teaching I have heard the good, the bad, and the ugly of statistics, but mostly the bad and ugly—"I hate stats", "I need it to graduate, ugh!", or "I'd rather be dead". Well, the reality of the matter is that I have spent a portion of my time in academia creating this textbook to assist students, faculty and any researcher to understand how to write and interpret the numbers, especially from an alien output. This book is basically a humble response for all those students who struggled with statistical interpretations and understanding SPSS outputs and analyses. This book should ease the process of data analysis and show you the countless ways to have fun with data, once you understand the test and the rules that guide it. This book is a unique creation and is based on my own experiences in the classroom as a student, as well as a professor. Many years ago, when I took undergraduate and even graduate statistics, it was simply a core course requirement that must be taken to fulfill my BA, MA, or PhD degree requirements. I, like others, did not embrace it well and enrolled because I had to. When I had questions about things like levels of measurement or coding or data modification, the TA would say, "it's in the book". It was never in the book or nicely laid out or simply placed where we could navigate a simple answer. Finding an answer always seemed so very complicated to find or locate. My goal for this book was to create a text for the super popular core statistical tests used in the social sciences with all the answers we look for as we analyze data, especially what to include or exclude and what conclusions can or should be made without being overzealous.

Statistics courses at any institution with any professor, remains challenging and anxiety provoking for many. "Doing numbers" is difficult, especially for students housed in the Social Sciences. In my undergraduate and graduate years, I always searched for a book that had basic answers to my questions. Nothing complicated, but some reference guide that told me the statistical recipes of analyzing data—like what level of measurement is needed for this test? Many times, I was directed to the textbook which didn't not simply tell me in black and white. I always detested when they said, "read the book". What book? I have read the book, and it doesn't say. Honestly, your experiences speak volumes when "doing statistics". Books don't have all the answers. The struggles are real and with the advent of technology, many software programs came to the forefront to aid the learning process for students and faculty. Today, these software technologies, like the Statistical Package for the Social Sciences (SPSS), have dominated the world of data analyses. I have taught statistics as a Teaching Assistant, Sessional Instructor, Assistant Professor and currently an Associate Professor at various institutions and the issues related to this type of course are classic. However, the journey that was most pivotal in my tenure was my time at Ontario

Tech University (previously known as University of Ontario Institute of Technology or UOIT). Here, I became an invested professor in the subject matter who was completely devoted to the task at hand: teaching stats and helping them understand the subject-matter which students could care less about. My goal from Day 1 in the lecture room was to deliver my lecture using a meaningful method and strategy; a method that helped students "get it" and understand that in working with numbers, there is an "art to the interpretation"-a method to the madness. Finding textbooks that spoke directly to the interpretations of outputs were rare. In the statistical textbook space, most books concentrated on explaining the various tests, providing umpteen formulas and provide homework exercises. There were very little books that discussed how to interpret SPSS outputs and discuss the key message from the tables technically and substantively.

My courses on data analysis taught students exactly this process. I would provide the theory of the test, the how, what, why, when and where, assumptions and/or requirements of each test, alongside the formula, run SPSS, get an output and spend the rest of the class helping students understand the interpretive workflows so the writing at the end of it all made sense quantitatively and for the layman's person. A lot of trees were cut and lots of paper trails of countless outputs for my students to engage in statistical interpretation. Over time, my office became a space for all types of SPSS outputs with interpretations and desks were overflowing with SPSS outputs and interpretations. Many students over the years recommended I convert my teaching or lecture style to a solid book, which students can use a reference guide. The idea brewed and brewed and finally, it has developed into this textbook.

The main objective of this book is to break down the statistical process in a systematic manner. Students need to understand in detail what each statistical test brings to the table and understand the core requirements for each statistical test. Each chapter informs the student of the test and tells them exactly why they would use this specific test and in what circumstances and with what kinds of data. This is so important. Understanding the basic principles of quantitative research, variables, levels of measurement are integral knowledge prior to running tests. Any student can memorize a formula, but to understand the exact nature of statistical test is another type of learning. In this way, students will appreciate the statistical methods and approaches and utilize these statistical tools and tests in a proper and profound manner. They will understand that the background score of each test and run according to the data requirements.

Little did I know then that statistics would become an integral part of my being, as an academic and in my daily activities. This book will be of great service to many stats students, undergraduate and graduate, as well as faculty and researchers, because it simply provides the necessary tools for moving forward in their statistical pathways to understanding how to analyze basic outputs of SPSS, both technically and substantively. Essentially, it is a simple book that provides valuable information for core statistical tests, that my own experiences in academia have taught me. I am forever grateful to my students who have constantly discussed things that work for them and things that do not in my statistic classes. This book is a product of all their questions, concerns, and comments throughout my teaching. It is hoped this book truly holds the essential answers to questions each student taking stats has and becomes a reference guide when writing numbers and doing some hardcore quantitative storytelling—this book will most definitely be *"statistically significant"* ($p < 0.05$) in the lives of those individuals doing statistical interpretations and data reporting.

Summary of Statistical Tests Discussed in This Book

I Descriptive Statistics

That branch of statistics that describes sample data by reducing it to a single number. These statistics organize and present data in a convenient summary (i.e., data reduction) format by providing trends and patterns of the data.

Univariate (One Variable) Analyses

1. Frequencies give the trends and patterns of the data by providing counts and valid percentages.

Independent of levels of measurement (i.e., can be run on any level of measurement); they tell you the observed count/frequency of each respondent in a specific response attribute and also provide you with the corresponding valid percentages.

2. Measures of central tendency summarizes data into the most typical, central or representative value of a distribution. Dependent on levels of measurement

Mode is the value that occurs most frequently in any distribution
Used mostly with **nominal** data
Median represents the exact centre of a distribution of scores.
Used mostly with **ordinal** data, but can also be run on interval-ratio data
Mean represents the arithmetic average of a distribution.
Used with **interval-ratio** data

3. Measures of dispersion or **variance** indicates the amount of heterogeneity or variation within a distribution of scores for **interval-ratio** data. The following are the most common ones used:

Variance represents the amount of variance in a sample.
Standard deviation represents how much a score deviates from the mean.

You discuss **skewness** and **kurtosis** of histograms for all interval-ratio data.

4. Visual representations or **Types of graphs**

Bar chart with percentages for **nominal** or **ordinal** data
Histogram with Normal curve for **interval-ratio** data or its equivalent

Bivariate and Multivariate Descriptive Analyses

1. Zero-Order crosstabulation, Measures of association, and **Chi-Square** tells the researcher whether a relationship between the IV (x) and DV (y) exists (e.g., Are stress levels related to suicide attempts?). The measure of association determines the strength and direction of the relationship, while chi-square examines statistical independence (not significant, $p > 0.05$) or statistical dependence (significant, $p < 0.05$); reject H_o or fail to reject.

(A)   A smaller crosstab is easier to analyze; it's simple
(B)   Any cell with less than five cases in it is not valid
(C)   (recoding is a possible solution)
(D)   Used with nominal and ordinal data
(E)   Phi, Contingency Coefficient C, Cramer's V, Lambda=NOMINAL
(F)   Gamma, Yules Q, Somer's d, Tau-b=ORDINAL
(G)   Anything close to 0 is considered weak; anything near 1.00 is strongly correlated.

2. Elaborated crosstabulation, Measures of association, and Chi-Square (one IV and one DV + control variable) tells the researcher whether a third unseen factor/variable is influencing the original zero-order cross-tabulation.

II Inferential Statistics

That branch of statistics that allows quantitative researchers to make inferences (i.e., generalize) from a sample to a population. Based on hypothesis testing and predictions.

Hypothesis Testing Techniques Using Comparison of Means Tests

1. Independent Samples t-test tells the researcher if there is a mean difference between two groups.

Used with a dichotomous IV and interval-ratio DV.
Be sure to state a null and alternative hypothesis where needed.
Statistics to be analyzed: Averages of both groups; Levene's Test for Equality of Variances; t-statistic and corresponding statistical significance. If $p < 0.05$ we reject H_o.

2. One Way ANOVA involves analysis of one IV with more than two groups/response attributes and one interval-ratio dependent variable.

Statistics to be analyzed: Descriptive statistics, especially averages of each group with respect to the grand mean; the upper and lower bounds; minimum and maximum values; the F-statistic and corresponding significance; post hoc test (Bonferroni). If $p < 0.05$ we reject H_o.

3. Two Way ANOVA/Factorial ANOVA involves the analysis of two or more IVs that are categorical and one interval-ratio DV. There *must* be two or more levels for each nominal/ordinal IV and 1 interval-ratio dependent variable.

Statistics to be Analyzed: sample sizes of each response attribute for each IV; the averages of each IV on the DV; Levene's Test for Equality of Variances; main effects;

interaction effect; partial eta squared and overall R-squared for model; post hoc tests (Bonferroni). If $p < 0.05$ we reject H_o.

4. ANCOVA or Analysis of covariance is an advanced multivariate ANOVA technique with more than two groups' averages being compared on a particular DV. It is a statistical analysis that combines components of ANOVA and regression. It basically evaluates whether population means of a DV are equal or the same across different levels of a categorical IV, while controlling for the effects of a continuous interval-ratio variable-covariates. ANCOVA lets researchers to control for effects of one or more control variable or covariate.

At least one IV or factor: One nominal or ordinal categorical or discrete variable with more than two response attributes; you can always add a second factor or IV.
Dependent Variable: One interval-ratio continuous or scaled.
Covariate: Ideally interval-ratio or its equivalent; but can be nominal or ordinal.

5. MANOVA and MANCOVA is a comparison of means hypothesis testing technique which is simply an extension of the univariate analysis of variance. In MANOVA, the analysis extends to multiple DVs. Thus, there are multiple measurable social phenomenon that can be tested through hypothesis testing. Through examining patterns of average group differences, MANOVA provides great insight into increase complex relationships. Its core uniqueness lies in its ability to work with multiple DVs and understand how different groups or conditions diverge across the set of dependent variables or outcomes. MANOVA creates a holistic analysis of the relationships in question and makes efficient use of data. MANCOVA, on the other hand, an extension of MANOVA introduces one or more interval-ratio continuous covariates to the model of "average differences" on sets of DVs. The key goal is to assesses group differences in the multivariate means of dependent variables while controlling for the effects of covariates. This statistical analysis allows to test for average group differences while accounting for the effect of covariates.

Independent variable: Requires two or more categorical **nominal or ordinal IVs** or Factors with two levels or more or response attributes
Dependent variable: Two multiple interval-ratio continuous or scaled variables with a shared conceptual meaning
Covariate(s): One to two interval-ratio continuous or scaled variables with a significant association with the DV

Association and Predictions with Interval-Ratio Continuous Variables

1. Pearson Product Moment Correlation (r) is a measure of association for interval-ratio data; it tells the researcher to which extent x varies with y by giving the magnitude or strength (#) and direction (+ve/-ve) of the relationship. A correlation is usually run prior to running a regression analysis to see how x and y co-vary. *Correlation does not indicate causation.*

Used with Interval-Ratio (or like-dummy coded 0–1) data
Analyze the #, sign, and significance. Anything close to 0 is considered weak; anything near 1.00 is strongly correlated.
If $p < 0.05$ we reject H_o.

2. Linear multiple regression/Ordinary least squares method (OLS) (Bivariate or Multiple), a very sophisticated test which puts in predictor variables *simultaneously* into the model and tells the researcher the change in Y (i.e., the dependent variable) that is associated with a change in X (i.e., independent variable). It is designed to help predict the most likely value for the other variable based on available information. In the model summary, the R^2 (coefficient of determination) provides information on the fit of the model by assessing amount of explained variance by each predictor variable. A large R^2 value is what researchers strive for. Unstandardized beta coefficients and standardized beta coefficients tell us the story of the IV with respect to the dependent variable.

Statistics to be analyzed: Descriptive statistics, correlations with respect to DV; the model summary R-square; the regression coefficients, specifically Unstandardized b's and standardized BETA, t-values and corresponding significance; as well as Tolerance and VIF values; casewise diagnostics, Cook's and Mahal's Distance, if $p < 0.05$ we reject H_o.

3. Hierarchical or Incremental regression analysis or Block modelling takes in a conceptually specified model in steps/blocks by grouping variables together. Again, this is like linear multiple regression analysis; the key difference here is that variables are *not* put in simultaneously, but rather in a very theorized specific manner.

Statistics to be analyzed: descriptive statistics, correlations with respect to DV; the model summary R-square for each block as well as the change in R-square and its corresponding significance; the regression coefficients, specifically Unstandardized b's and standardized BETA, t-values and corresponding significance; as well as Tolerance and VIF values; casewise diagnostics, Cooks Distance or Mahal's Distance, if $p < 0.05$ we reject H_o.

4. Logistic regression is different and unique in so many ways, comparatively speaking. It predicts a binary DV outcome, such as yes or no. This type of regression predicts the DV by analyzing its relationship with an IV (Binary logistic regression); it is bivariate or multiple IVs (Multiple logistic regression). Many independent and dependent variables we want to understand do not occupy interval-ratio continuous status. Instead, they are in a binary, dichotomous, dummy coded state of 0 and 1.

III Data Modification and Statistical Manipulation of Data

1. Factor analysis is a *data reduction technique*. The main goal of factor analysis is to get at the *underlying structure* of data by allowing researchers to combine multiple measures/indicators/questions that measure almost the same social phenomenon. The underlying structure is referred to as a *latent construct* (i.e., underlying factor). It may be
   Exploratory or confirmatory in nature. The variables should assume nominal or ordinal properties and conceptually related. All variables or measures should all measure ONE social phenomenon and response attributes should be directionally similar with an adequate sample size (i.e., at least 50 cases per variable)Statistics to Analyze: descriptive statistics; KMO and Bartlett's Test of Sphericity; Total Variance Explained and Eigen Values; Rotated "varimax" solution; anything with a factor loading > 0.40 will load on a specific factor.

2. Reliability analysis is a reliability test for multiple measures; it tests to see how well measures hang together; are they measuring what they should be measuring; is the internal consistency of measures good or not, so that standardized scales can be built.

3. Recoding & Dummy Coding is the process of data manipulation that allows for the collapsing of response attributes that can easily group together. Recoding is done for two reasons: 1. To increase cell counts; 2. To change the nature or type of variable. Example non-dichotomous to dichotomous; dummy coding, assigning values of 0 and 1 to the response attributes to create interval-ratio like variables or its equivalent.

4. Scaling is the process of constructing interval-ratio like continuous data from nominal or ordinal type data. A standardized scale is built by summating the variables or measures of interest and divided by the # of variables (i.e., v1 + v2 + v3 + v4 + v5)/5.

5. Sample manipulation and analysis can be done using Select Cases and Split File Commands in SPSS.

Part 1

Introduction to Quantitative Research Methods, Data Modification and Descriptive Statistics

The first portion of this book explores the essentials of quantitative research methods, data modification, and descriptive statistics. Upon reading and reviewing this chapter you should be able to develop an appreciation for quantitative research methods and how they deviate from qualitative research methods, as well as understand how data modification is utilized in statistics, and finally fully appreciate the breadth of descriptive statistics, both univariate and bivariate analyses, as well as multivariate analyses. This discussion is important, as it lays the groundwork for a larger discussion around more sophisticated and complex statistics. Univariate Statistics, Zero-order Crosstabulations and Elaborated Crosstabs are the topics that are examined at length and explored. These are the most common descriptive statistical tests used in the social sciences and humanities.

1 Understanding Quantitative Research as a Precursor to the Statistical Pathways

Introduction to Quantitative Research in the Social Sciences & Humanities

Most people in the social sciences cringe when they hear about statistics or quantitative methods or data analysis, advanced or not. There is nothing exciting or intriguing about quantitative analyses or the numbers game. Most often, anything quantitative or number related creates much anxiety and butterflies. There is an overwhelming cowardly response when enrolled in statistics or data analysis or quantitative research methods courses. There are many 'academic' names to this type of [core] course. Quantitative research methods is not easy and when coupled with statistical analyses it becomes quite overwhelming and a daunting task for most of us. We do it because we must, not because we want to. Whether this course is taken in second, third, fourth year or even graduate school, the anticipation of 'doing statistics' becomes exponentially intimidating in so many ways. The bad news is that most of you will have to engage in quantitative research in academia or in work-related spaces. There really is no running away from numerical analyses and data crunching. It is everywhere and with the advent of Artificial Intelligence (AI) and data science, statistics is becoming a harsh reality. The good news is that if we think about the value quantitative methods and analyses bring to the table, we may learn to appreciate their methodological promise and potential. Our daily lives are consumed with statistics. Sometimes, statistics are seen at political polls or while shopping or while engaging in event planning or counting COVID-19 pandemic numbers. The numbers game is here to stay and it is always telling a story of sorts. Thus, it becomes important to put some effort into understanding data analysis and the beauty each statistical test, basic or advanced, has to offer. Each test discussed here has its own merit and brings its own unique contribution. The significant part to understand is how to use which test, and when to use it. Additionally, you need to know how to interpret the numbers, results and make sound conclusions about your quantitative data.

Quantitative research methods, sometimes known as data analysis or simply statistics, governs our social and methodological inquiry process and thus must be understood. The social sciences and humanities are thought-provoking and curious disciplines which are inevitably bound to investigative and social inquiry of the world around us. The social world awaits discovery either objectively or subjectively, methodologically speaking. Regardless, this discipline methodically quantifies or qualitatively takes day-to-day observations and through scientific rigor creates reliable and valid observations and findings, results, or outcomes. At the end of the day, a strong and systematic methodological stance allows one to create great data and research. The methodological agenda of the social sciences is rooted in two fundamental methodological approaches of social inquiry: Quantitative and Qualitative. Both are similar in that they are two [unique] ways to navigate the social world – however, how they go about doing this is very different.

DOI: 10.4324/9781003215691-2

The quantitative footprint and objectives, which is the focal point of this book, are unique and require some discussion.

To begin, *Quantitative methodology* is defined as a positivist, systematic, methodological tool of inquiry that encompasses research techniques that involve and include the numeric representation of numbers through data, manipulation, and analyses through processes of large randomized probabilistic sampling methods, like simple random sampling, systematic sampling, stratified sampling, and multi-stage cluster sampling. Its scientific method stems from a deductive, linear, and macro-oriented pathway to investigate the empirical social world. The social reality of the quantitative researcher is driven by an objective viewpoint, in which reality is 'out there' waiting to be discovered, unlike the subjective, empathy-driven or *verstehen* qualitative researcher. Quantitative research is used across many disciplines and thus has become interdisciplinary in nature. The social sciences, natural sciences, business, economics, and engineering frequently engage in these methods. From its objective measurement to research design, to data collection, variables, statistical analysis, and quantitative instruments it creates a world of endless possibilities. Data collection in quantitative methods can be done using survey research or questionnaires (in-person paper pencil, online), experiments, content analysis or secondary data analysis.

In contrast, a qualitative researcher who is interpretive in their methodological thought, utilizes a micro-oriented approach that is inductively achieved through non-randomized sampling methods (i.e., convenience sampling, snowball sampling, purposive sampling, etc.) and through non-numerical accounts (i.e., narratives, phrases, themes, etc.) to better understand the true, lived experiences of participants. Here, large samples are not the goal but, rather, quality data and outcomes. The focus is on a unique perspective and voice that is explored through verbatim expressions and thick descriptive accounts of their definition of the situation or, simply put, their social reality. The space of qualitative research focuses on the nature of data, research design, its purpose, data collection tools, iterative processes of data analyses of transcripts and narratives, as well as its emic perspective of the insider's perspective vs. the outsider's perspective. How one goes about investigating the social world quantitatively or qualitatively depends on many things, like feasibility, cost, time and whether the sample is accessible, as well as achievable or not. Additionally, the most important question to ascertain is: does my research agenda and question(s) pair well with the methodological research design and theoretical framework (i.e., social explanations)? (Denzin & Lincoln, 2000; Creswell & Creswell, 2017; Neuman, 2013; Stockemer, 2019). Both methods are based on ethical considerations and Research Ethics Board (REB) or Institutional Review Board (IRB) approval.

Understanding that quantitative methodology does not encompass a fluid, flexible and reflexive mindset is an important undertaking, as it sets the stage for how the 'quantitative expectations and assumptions' play out in the real world. The world of numbers in the social sciences is rooted in this philosophical approach. The epistemological and ontological principles of quantitative research drive the numbers game of Statistics. The quantitative research cycle follows a rigid research cycle and goes through a step-by-step process to get to a result. Most often, it begins with selecting a topic that you are interested in, then a review of the past scholarship to explore what others have written or not about the subject matter, selection of a theory or theories to understand the social explanations that further explain the topic, concepts and variables in question, followed by building a broad research question, with clearly defined null and research hypotheses that are measurable and testable, engaging in survey research design and being cognizant of your statistical analyses, as well as levels of measurement needed to analyze data, getting research and ethics board approval, going into the field and collecting the survey data, engaging in data cleansing, analyzing and interpretation(s) and lastly disseminating findings through articles, conferences, and workshops.

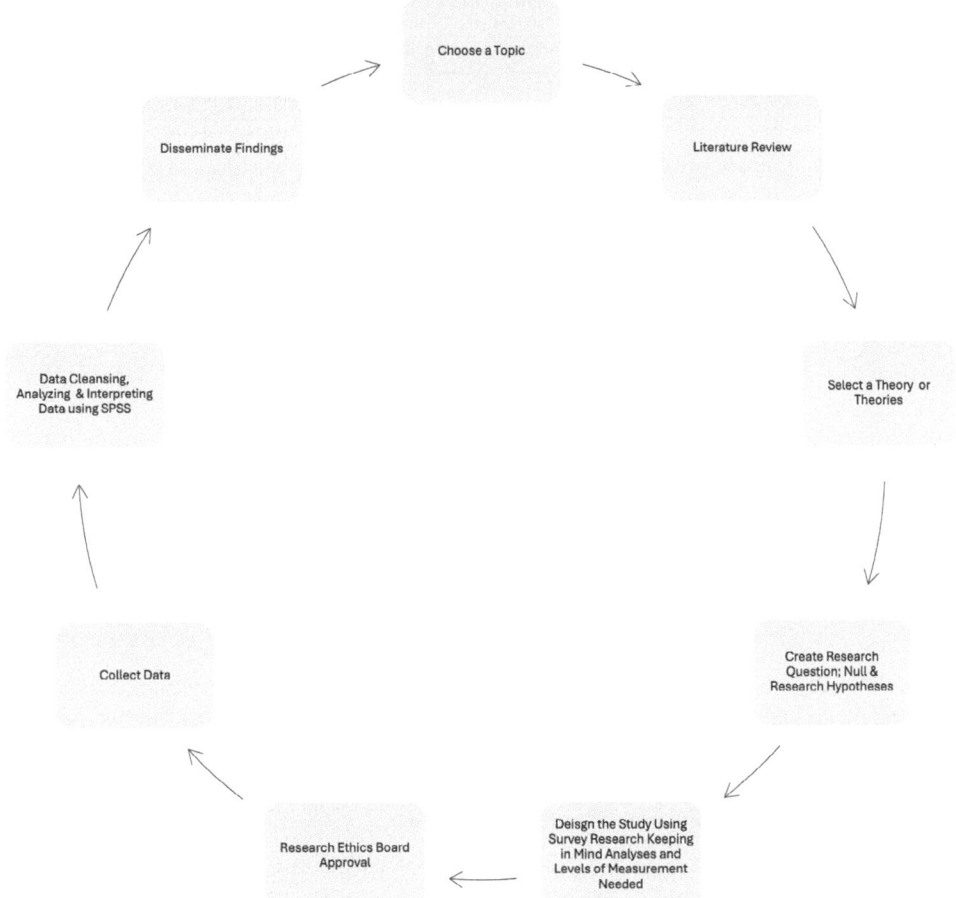

Figure 1.1 Quantitative Research Cycle Process. A (re)iterative process that most often follows these steps, starting from Choose a Topic and ending with Dissemination of Findings

Quantitative research and analyses, unlike its qualitative counterpart, follows a rigorous method of processes and procedures that should be followed, at its best, to have reliable and valid outcomes. The first step of any investigation of the social world is to find a suitable topic that interests you and drives a research interest and question to ponder. It is important to state at this point that many research designs combine both quantitative and qualitative methodologies. This is referred to as *triangulation of methods* or a *mixed method* design (Creswell & Creswell, 2017). In this way, both methods complement the research agenda and create great research in the process. The intricacies of objective and subjective data collection beautifully capture the essence of research depth.

Another point worth noting is how participants are handled with respect to time in quantitative and qualitative research methods. There are two types of studies: 1. Cohort Study; 2. Longitudinal Study. A cohort study is one in which participants share a common experience or characteristic, like year of birth or health conditions or career. In this way, they are followed over time to observe certain outcomes. A longitudinal study is slightly different. In this type of study, the same group of participants is repeatedly observed or measured at multiple points in time. Here, researchers observe changes within the same group over time. The research objectives and agenda define what type of study will take place. Also, time and cost issues. Feasibility of study is most important.

Summary of the Variations Seen in Quantitative and Qualitative Methodologies

	Quantitative	Qualitative
Meaning and objectives	Manipulation of data and numerical analyses using statistics, descriptive or inferential	Non-numerical analyses transcripts to extract themes and narratives through verbatim words
Methodological approach to understand reality	Positivism: Social reality is objective waiting to be discovered; begins with larger structures (macro)	Interpretivism: Social reality is subjective; Empathy-driven; micro-oriented; begins with the individual (micro)
Research method	Quantitative, deductive, linear, and rigid	Qualitative, inductive, non-linear, flexible, and grounded theory approach
Data collection	Surveys, experiments, content Analysis, secondary data	Interviews, participant observation, ethnographies, narratives, focus groups
Analyses	Statistical analyses using data analysis, most basic to advance to test our hypotheses	Thematic coding and analyses utilizing thick descriptive accounts to unveil their definition of the situation and social realities; give voice to participant
Sample and sampling	Randomized samples using Equal selection probability selection methods (EPSeM); based on Central limit theorem and the Law of large numbers where generalizability of findings is key with normalized data	Non-probabilistic methods; non-randomized; small samples where goal is not generalizability of findings but findings a 'unique' case or cases or themes
Reliability and validity	Measures and analyses should be reliable, valid, and tested using statistical tests like a reliability analysis	Giving a unique voice and perspective to participants should be authentic via empathetic understanding and capturing the true voice of what was said verbatim
Outcome(s)	Statistical reporting that should generalize findings from a sample to a population	Explores narratives, themes, and quotes case by case; each case is unique; to capture authentic voices and not make grand inferences; gain a unique perspective

Descriptive and Inferential Statistics

Statistics, in very simple terms, is defined as the numerical assessment, manipulation and analyses of numerical data that is collected and categorized unequivocally based on their levels of measurement (i.e., nominal ordinal, interval, and ratio). Sir Francis Galton was one of the first pioneers of statistical thinking and conceptualization. He was one of the first individuals to apply statistical analyses to individual differences and intelligence in genetic studies. We have him to thank for the data analysis or statistic courses you are in. Fundamentally, there are two main branches of statistics:

Descriptive Statistics

Descriptive statistics is defined as that branch of statistics that describe the 'trends and patterns' of the raw data through a process of data reduction. It simply summarizes and presents data in meaningful analytics. These statistics can be univariate, bivariate, or multivariate: univariate meaning single variable analysis; bivariate meaning two variable analysis (i.e., cross-classifying), specifically dealing with an independent variable and dependent variable; and multivariate, adding a third unseen variable, known as the control variable, to the model. Descriptive statistics are limited in their analytical capabilities and specific to the sample being analyzed. Here, generalizations from a sample to a population are unheard of. With research, Descriptive statistics is often used to provide key demographics of a sample and build a sample profile.

The most used Descriptive statistics include the following univariate and bivariate statistical tests:

a   Frequencies
b   Measures of central tendency: mode, median, and mean
c   Measures of dispersion or variation or heterogeneity: range, variance, standard deviation
d   Visual representations, like Bar charts and Histograms
e   Zero-order crosstabulations and First order or Elaborated crosstabulations

Descriptive statistics assess trends and patterns of variables and distributions by levels of measurement. They may be univariate, bivariate, or multivariate in nature. The goal here is not to make inferences from a sample to a population but rather use numerical representation to describe the social world through storytelling of numbers. This number can be a count, a percentage, a common score, mid-point, average, spread of scores, or cross-classification of variables. These statistics, although not so powerful, provide very critical information on the nature of distributions, the typical average or score of a distribution, and the spread. Prior to running advanced statistical tests, having information about the distributions is imperative. Decisions regarding data modification are made at this point as well.

Univariate	Bivariate	Multivariate
Single variable analysis	Two variable analysis	> two variable analysis
Frequencies (all levels of measurement)	Zero-order crosstabulation (nominal, ordinal)	Elaborated Crosstabulation (nominal, ordinal)
Measures of central tendency • mode (nominal) • median (ordinal, interval-ratio) • mean (interval-ratio)	Sample statistic: Chi-Square	Sample statistic: Chi-Square
Measures of dispersion • range (ordinal) • variance, standard deviation (interval-ratio)	Measure of association	Measure of association
Graphs • bar chart with %ages (nominal, ordinal) • histogram with normal curve (interval-ratio continuous)		

Advantages of Using Descriptive Statistics

Parsimonious	Simple and easy to understand. Present data in simple ways
Data reduction or summarization	Condenses large amounts of data into few key statistics, like mode, median, and mean
Central tendency and variation	Provides the typical score of distribution and spread of scores
Data distribution	Provides visual depictions of distribution of data
Identifies outliers or extreme scores	Identifies outliers and/or extreme scores that may impact distributions positively or negatively
Comparisons	Allows to compare different groups and datasets
Decision-making	Allows to make informed decisions about future and more advanced statistical analyses

Inferential Statistics

Inferential statistics is a more complex, powerful, and advanced branch of statistics that is defined as that level of statistics that allows for inferences to be made from a sample to a population from the raw data. We cannot study everyone-the population. There is not enough money or time to do so. Thus, we work with samples, subset of populations that provide sample data to estimate characteristics of the population parameters. Here, sample size matters. The larger the sample size, the greater our confidence levels are. The beating heart of Inferential Statistics is the *Central limit theorem* and the normal curve or Bell curve. This theorem unapologetically states that if you have a population with mean μ and standard deviation σ and take large random samples from the population, then the distribution of the sample means will be approximate the normal distribution. The sample means approach a normal distribution with a large N (Meyers, Gamst, & Guarino, 2016. The normal curve is an ideal symmetrical bell-shaped distribution and unimodal. Here, the mode, mean, and median values at the center of the distribution are equal. There are no extreme scores or outliers and the measure of dispersion, standard deviation determines the size of the Bell curve. Obviously, larger values of standard deviation lead to greater spread of scores or increased variation of data. The rule of thumb is that the normal curve is distributed into three ways or often known as the 68–95–99.7% rule, with data falling within 1 standard deviation, 2 standard deviations or 3 standard deviations, respectively. Most sophisticated statistical tests assume that data follow the path of the normal curve, hence making this a key reference point in statistics, especially inferential statistics.

Here, generalizability of data or findings lies at the core, providing that Equal probability selection methods (EPSeM) are utilized and ultimately the sampling distribution equates the population parameters such that inferences from a sample to a population can be readily made, provided all statistical test assumptions and requirements are met. Some common probabilistic random sampling methods are: Simple random sampling, Systematic sampling, Stratified sampling, and Multi-stage cluster sampling. It is important to note that in real life sometimes quantitative researchers utilize non-probabilistic methods as well and engage in Convenience sampling.

Core inferential statistics include, but are not limited to, independent samples t-test, One-way ANOVA, Two-way factorial ANOVA, ANCOVA, MANOVA, MANCOVA, Pearson Product Moment Correlation, Partial correlation, Multiple linear regression, Hierarchical or Incremental regression and Logistic regression (Bachman, Paternoster

& Wilson, 2022). There are many more data modifications done in this branch of statistics because inferential statistics have many assumptions tied to them regarding the data that must be met.

Even though differences are discussed amongst these two branches, it is also important to pay attention to how both branches have complementary roles in statistics. They basically work in conjunction with one another. Analyzing trends and patterns prior to running any inference-based test is critical to any analysis we undertake and we must work in sequence. Descriptive statistics, although heavily exploratory, describe initial results and then, based on those results, make informed decisions accordingly about the inferential testing that should be done – whether the end-goal is average differences, relationships, or predictions. In keeping with the research objectives of the research, informed decisions are made of what tests to utilize to maximize that analytical quality of the data. All statistical test assumptions and/or requirements must be fulfilled to their best to ensure that any statistical errors are minimized.

Advantage of Using Inferential Statistics

Generalizability	Allows to generalize from a sample to a population
Hypothesis testing	Allows to formally test a research hypothesis and examine if a relationship exists or average differences exist or whether certain predictors are significant in predicting a particular outcome
Decision making	Makes informed decisions and meaningful conclusions by understanding the statistically significant outcomes of any hypothesis testing relationship
Associations and predictions	Provides information about the association of variables and predictability of variables
Confidence intervals	Provides confidence intervals and p-values that allow us to reject or fail to reject a null hypothesis

Concepts, Variables, and Levels of Measurement

Quantitative researchers most often are heavily engaged in survey research, paper-pencil method or online to collect data. Survey research is by far the most popular data collection technique amongst quantitative methodologists. This data can be primary or secondary data. *Primary data collection* occurs when data is collected by the researcher themselves. The researcher collects the data firsthand and analyzes it as well, whereas *secondary data collection* is data collected by another person or researcher. Here, the data is readily available to the public for (re) analysis of it. All the questions that are part of the survey instrument denote a concept. Each concept is carefully *operationalized* to the terms of the research. Defining terms are critical to the quantitative research process. Each abstract or concrete concept requires a definition that matches with the research objectives and agenda. Once defined, each carefully created question must fully embrace the definition and cover all aspects. Only then *variables* are generated. Variables are the basic unit of statistics. In quantitative methods, there are many types of variables. It is important to understand that each question on a survey denotes a variable (Neuman, 2013).

The Characteristics of a Survey Question and Variable

Concept: Criminal Record
Operationalization: Respondent has a criminal record or not.
Survey Question: In your lifetime, have you ever had a criminal record? [CRMREC]
[0] No
[1] Yes

The 'no' and 'yes' are the response attributes and the 0 and 1 represent the coding for that question. This question now has become a variable in a dataset.

Quantitative research methods are driven by variables. They are an essential part of research design and analysis. It is important to understand the differences amongst variables. That is the first step of analyzing data and building hypotheses. The most popular variables are referred to as 'independent' and 'dependent' variables. The independent variable (IV), often denoted as X, is best known as the causal factor, those that come first in time, and is manipulated or controlled by the researcher. In research that examines one's criminal record and employment, employment would be identified as the IV. Working full-time, part-time, or not employed. The dependent variables (DV) often denoted as Y, is known as the effect, outcome, or criterion variable. It is identified as what is being measured or analyzed in any research study. In the criminal record and employment example, criminal record is the DV. The association between the IV and DV is critical to any statistical analysis. The thrust of any statistical test is to clearly determine whether changes in the IV cause, or are associated with changes, in the DV. Cause and effect is at the front end of statistics and the definitive end game. Here, careful thought and consideration must be given to any research design trying to establish cause and effect.

Try to identify the IV and DV in these examples:

- Level of education and delinquency rates
- IPV victim and time in Canada
- Substance use and violence in the home
- Cyberbullying and living arrangements
- Social media and self esteem
- Anxiety and social maladjustment
- Political party affiliation and gun laws
- Years of education and current salary
- Race and criminal record

Often, these variables are visualized like a cause-and-effect relationship (see Figure 1.2):

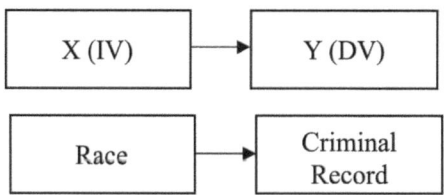

Figure 1.2 Arrow Diagram or Causal Model of an IV and DV in a Quantitative Relationship

There are also other variables worth mentioning in quantitative methods – like intervening, and control variables seen in models. An 'intervening variable', also referred to as a mediating variable, is a variable that is positioned between the IV and DV and intervenes the relationship. It is a variable that influences or impacts the IV and DV both and creates a nuanced causal pathway. For example, researchers may be interested in socioeconomic status (IV) and higher rates of criminal behavior (DV). Individuals who do not come from affluent backgrounds may engage in criminal activity. However, a mediating or intervening variable, like lower educational attainment, may also contribute to the increased criminal behavior.

Alternatively, a *control variable*, also known as a *covariate* (in other statistical tests), often denoted as 'Z' is an unseen third variable that tests for *spuriousness* of an original x and y relationship. By holding other variables or factors constant, it ensures that there is no other variable that may impact or influence the original x and y relationship. Basically, it influences the outcomes of a relationship being tested and provides reassurance on the unique contribution of the IV to the DV. Addition of control variables to a model can alter the outcomes significantly and such findings can provide further insights into the original relationship. Demographic type variables – like age, gender, race, education, marital status, to name a few – are often used as potential control variables to see and investigate further how the relationship is modified or not. However, sometimes more specific variables relevant to the topic may be used. For example, if I am testing a relationship between survivors of domestic violence and race, I can control for family support variables or social support variables. I could also control for other demographic variables. The use of a control variable really works towards your theoretical or social explanations of the topic being researched.

Knowing the functionality and versatility of types of variables used in quantitative analyses is important to hypothesis building and relationship testing. These are the main variables used in statistical analyses. Knowing the role of each of these variables becomes critical to moving forward in statistical data analysis. Being able to identify, with logic and reason, which variable is what provides better understanding of the analyses going forward. Not having the exact knowledge of the type and nature of variables early on, may create major setbacks moving forward, especially as it relates to levels of measurement in statistical analyses.

In quantitative methodology, each variable goes through a variable classification process by levels of measurement, namely *nominal, ordinal, interval*, and *ratio* (NOIR). These are the key levels of measurement, and each conveys its own story about the nature of the data. No two levels of measurement are equal and move in a hierarchy of order of lowest level of measurement (nominal) to highest level of measurement (interval-ratio). Levels of measurement are critical to any statistical analysis because they undoubtedly inform our statistical pathways and the tests we can run. Mostly, each statistical test has a level of measurement for each variable, and this must be validated for reliability purposes. In the statistical thinking and strategies that you utilize, it is imperative to be able to take a survey question or variable and assign an appropriate level of measurement to it. Training yourself to do this is the initial step and very much critical to learning statistics. Each level of measurement comes with its own definitions and categorizations. For example, nominal variables are categorical with mutually exhaustive response attributes. Variables like gender, race, marital status, religion, etc. fall in this category. There is no order or numeric response. The response attributes are simple categories (i.e., words, like yes or no). Nominal variables are simple and easy to work with. They are the least sophisticated and present as one of the simplest variables. Ordinal variables have some inherent order or ranking of response attributes. They are often represented as a Likert scale continuum, like strongly agree, agree, disagree, and strongly disagree. These variables can be classified as categorical or discrete. Categorical variables can be ordered and have non-numeric response attributes; discrete variables are numeric with ranges of equal intervals, when possible. Variables that measure rank or status are often labelled as ordinal. Interval variables are continuous numeric variables with order and equal intervals between each response attribute. They symbolize the highest level of measurement. Here, there is no true zero point, rather an arbitrary one. Thus, a score of 0 does not inherently mean 0. A variable like temperature, an aptitude test score, or an intelligence test score, etc. Ratio variables, like interval variables, are continuous numeric variable with order and equal intervals between each response attribute, however they contain a true zero point. A score of 0 means 0. For example, # of times arrested, annual income, # of missed calls, # of times diagnosed with a personality disorder, years of education completed, etc. As you navigate the levels of measurement space in upper year statistic courses, it becomes clear that interval and ratio become one, namely interval-ratio.

NOIR Levels of Measurement Classification from Lowest to Highest

Level of measurement	Description	Example
Nominal	A categorical variable that is non-numerical. Nominal variables designate categories that are qualitatively different from one another and mutually exclusive. There is no order to these response attributes. They compromise of the lowest level of measurement	Gender, race, religion, marital status, and political affiliation
Ordinal	A categorical or discrete variable that is non-numerical and numerical (with ranges), respectively. These variables maintain some kind of order and follow a Likert scale	Professor ranks, employment category, and age in ranges
Interval	A continuous numeric variable that can be categorized, ranked, and has equal intervals, but has an arbitrary zero-point. The zero does not indicate complete absence of attributes being measured	Temperature, test scores, Weight, personality tests
Ratio	A continuous numeric variable that can be categorized, ranked, and has equal intervals, but has a true zero-point	# of kids, # of months on the job, salary, # of arrests

Activity Alert

List all four levels of measurement and the type of variable. Provide examples of each level of measurement.

As you move through the hierarchy of levels of measurement, you notice how statistical test sophistication varies with the level of measurement of variables. In bivariate and multivariate testing, most often, a combination of levels of measurement are working together, depending on which test is being run.

Hypotheses: Null versus Alternative or Research and Confidence Levels

Finally, these variables, through logic and reason, as well as academic scholarship, go through the process of conceptual specification and re-specification as hypotheses are created and tested. It's a process of statistical evolution and 'educated guesses'. Every research agenda has a broad research question that provides the general question of inquiry. For instance, does *juvenile delinquency vary by parental neglect?* In quantitative testing, specifically bivariate testing, there are two key hypotheses: the null and research hypotheses. The *Null hypothesis* (H_o) is an educated statement of no relationship, no difference, or no effect. The null hypothesis is the 'fall back' hypothesis or your baseline hypothesis. Alternatively, the *Research hypothesis* (H_1 or H_a) is an educated statement of what you as a researcher are hypothesizing that clearly says there is a relationship, difference, or effect. It is a claim that validates your research claims. It directly contradicts the null hypothesis.

An example of a Null hypothesis and Research hypothesis are:

H_o: There is no statistically significant relationship in parental neglect and juvenile delinquency.

H_1 or H_a: There is a statistically significant relationship in parental neglect and juvenile delinquency.

In statistics, based on the analytical results one is always in the position of rejection of the null hypothesis, partial rejection of the null hypothesis or fail to reject the null hypothesis. Rejection or partial rejection of the null occurs when there is an effect observed. The sample statistic allows us to reject or fail to reject the Null hypothesis. The sample statistic varies with each test. Each test has its designated test statistic. They commonly range from Chi-Square, t-value, F-statistic, r, but are not limited to this.

In statistics our rejection of the null hypothesis is based on three main *confidence levels*, 90%, 95% and 99%. These confidence levels speak to how confident we are about our results. Each confidence level has its corresponding alpha levels, 0.10 or 10%, 0.05 (5%), and 0.01 (1%). Thus, with a 95% confidence level, any researcher is willing to be wrong 5%. Any sample statistical value is deemed statistically significant if the probability of the calculated test statistic is less than an alpha of 0.05 ($p < 0.05$). If there is statistical significance, then we are in a position of rejecting or partially rejecting the null hypothesis. A 95% confidence level is most popular to use. It is the man in the middle, where there is not too much under confidence or over confidence. However, the choice of which confidence level to utilize depends largely on the sample size, sampling error and bias and any other research design flaws. At this point, we can even add control variables to the model to test for spurious relationships. However, if the $p > 0.05$ than the probability of the calculated test statistic is deemed insignificant, indicating no relationship or no effect. At this point, we fail to reject the null hypothesis and perhaps reconceptualize the relationship and test other variables against our dependent variable. Statistical significance is heavily linked to hypothesis testing and relationship-based testing. Small p-values indicate a statistically significant outcome, suggesting that results are unlikely due to random chance.

The Power of Technology Using Statistical Software: SPSS

In a technologically driven world, doing basic or advanced statistics using technology rather than traditional hand calculations creates a unique platform, especially for humanities and social science students that really do not like formulas. As social scientists, the technological software(s) used for statistics have overcome this fear and more and more students are open to data analyses for this very reason. The statistical package for the social sciences (SPSS) is one of the major platforms used to engage with data in a meaningful way and relatively easy manner. SPSS really streamlines data analyses and performs all the traditional mathematical calculations behind the scenes.

It all starts in the survey research phase. Once data is collected, the survey information, the data, are transferred to the SPSS data platform. Data entry takes place and variables are built, coded, labelled and created for further analysis and testing. The time taken in creating an accurate and properly defined and labelled data set speaks volumes during the analysis phase. Doing things methodically and systematically allows for great analysis and easily interpretable outputs. The task of building a data set is quite cumbersome and takes much time and planning. Once data entry is complete, you can begin analyzing the data by assessing and evaluating your test requirements and/or assumptions to create valid and reliable statistical analyses.

The statistical evolution towards a software like SPSS make things like Artificial Intelligence (AI) plausible. Today, AI is delivering real-time decision support systems for data scientists, as well as others. One of the essential foundations of AI is statistical methods that assist in development of models. With increased computing power, coupled with software sophistication, it is becoming easier for everyone to engage in active data analyses and provide reasonable outcomes about populations from various samples. The future of statistics is not without AI. Examination of trends and patterns, causal linkages, average differences, predictions, and making inferences all demand an advanced software, like SPSS.

I have used all free open-source secondary data that is readily available by SPSS or other data libraries. Most often secondary data is the option for most professors use. In this way, readily available data sets become the choice. Once the primary data is created or secondary data set is open, you can begin the process of statistical analyses and interpretation. The issue becomes how to read the outputs that are created. There is a method to do this, and it becomes imperative to not only know how to run SPSS, but it goes beyond this. You must know the following:

1   What does each statistical test offer and measure?
2   What are the core test requirements and/or assumptions for each test?
3   What levels of measurement are required to run the statistical test?

Final Thoughts

The social sciences and humanities are qualitatively and quantitatively methodologically driven and investigated accordingly. Without knowledge of research methods, research becomes tricky. This chapter focused on the latter, the quantitative paradigm and the core concepts that are aligned with quantitative methodological thought. All material discussed in this chapter covered the principles that lie at the heart of quantitative methodology, specifically statistics. This chapter has provided you with an overview of the quantitative methodological approach and the key characteristics that encompass this method. Descriptive and Inferential statistics, concepts, variables, and levels of measurement, as well as hypotheses, confidence levels and statistical significance were discussed. All these together encompass the key ingredients for quantitative research which eventually act as a precursor to the larger statistical pathways and conversations, especially with programs like SPSS, the statistical package for the social sciences. The nexus of technology and data analysis allows for sophisticated data analyses at all levels and is the future.

Keywords and Definitions

Quantitative methods	A positivist, systematic, methodological tool of inquiry that encompasses research techniques that involve and include the numeric representation of numbers through data, manipulation, and analyses through processes of large randomized probabilistic sampling methods.
Qualitative methods	A qualitative researcher utilizes a micro-oriented approach that is inductively achieved through non-randomized sampling methods (i.e., convenience sampling, snowball sampling, purposive sampling, etc.) and through non-numerical accounts (i.e., narratives, phrases, themes, etc.) to better understand the true lived experiences of participants.
Statistics	In very simple terms, statistics is defined as the numerical assessment, manipulation and analyses of numerical data that is collected and categorized unequivocally based on their levels of measurement (i.e., nominal ordinal, interval, and ratio). Sir Francis Galton was one of the first pioneers of statistical thinking and conceptualization. He was one of the first individuals to apply statistical analyses to individual differences and intelligence in genetic studies. Fundamentally, there are two main branches of Statistics.
Descriptive statistics	That branch of statistics that describes the 'trends and patterns' of the raw data through a process of data reduction. It simply summarizes and presents data in meaningful analytics. These statistics can be univariate, bivariate, or multivariate.
Inferential statistics	A more complex, powerful, and advanced branch of statistics that is defined as that level of statistics that allows for inferences to be made from a sample to a population from the raw data. We cannot study everyone – the whole population. There is not enough money or time to do so. Thus, we work with samples, subsets of populations, that provide sample data to estimate characteristics of the population parameters.
Generalizability of findings	A quantitative researcher, using inference-based statistics, ensures that results that are analyzed equal the population parameters and sample findings can be extended to the population at large.
Univariate statistics	Single variable analyses. One variable at a time is analyzed independent of others.
Bivariate statistics	Two variable analyses. Two variables are analyzed together, most often forming a dyad relationship of the independent and dependent variables.
Multivariate statistics	Multi-level analyses in which there are multiple IVs or DVs, and even control variables.
Primary data collection	Data is collected by the researcher themselves. The researcher collects the data firsthand and analyzes it as well, whereas secondary data collection is data collected by another person or researcher.

Secondary data collection	Data is collected by another person or researcher. Here, the data is readily available to the public for (re) analysis of it.
Operationalization	Defines concepts into the way it is being measured in the research and the way the researcher deems the concept to be measurable. It can be an abstract or concrete concept. Any survey question designed should match the operationalization of the term.
Equal probability selection methods (EPSeM)	A randomized method of gathering participants in quantitative research such that each person has an equal chance of being selected. Examples include simple random sampling, systematic sampling, stratified sampling, and multistage cluster sampling.
Independent variable	The causal factor or variable that is manipulated by the researcher. Most often comes first in time. Sometimes, referred to as predictor or factor. In more advanced techniques it is also referred to as the exogenous variable.
Dependent variable	The outcome or criterion variable. It is what is being measured and cannot be manipulated. This is referred to as the outcome or criterion variable. In more advanced techniques it is also referred to as the endogenous variable.
Control variable	A test for spurious relationships and a third unseen variable that is added to any bivariate relationship. Most, often these variable tests the original relationship of the IV and DV.
Level of measurement	How each variable is identified. Variables can be nominal, ordinal, interval, or ratio. The characteristics of the response attributes for each question or variable determines the level of measurement. These are important because they inform us as to which statistical test is plausible to run given the level of measurement properties.
Null hypothesis	The baseline hypothesis which suggests that there is no statistically significant relationship amongst the IV and DV.
Research hypothesis	An educated guess which suggests that there is a statistically significant relationship amongst the IV and DV.
Central limit theorem	The larger the sample size, the greater our confidence levels are. The beating heart of Inferential statistics is the Central Limit Theorem and the normal curve or Bell curve. This theorem unapologetically states that if you have a population with mean μ and standard deviation σ and take large random samples from the population, then the distribution of the sample means will be approximate the normal distribution.
Confidence level	The level in statistics our rejection of the null hypothesis is based on 3 main confidence levels, 90%, 95% and 99%. These confidence levels speak to how confident we are about our results. Each confidence level has its corresponding alpha levels: 0.10 or 10%, 0.05 (5%), and 0.01 (1%).

SPSS   The statistical software package of the social sciences that allows you to analyze large amounts of data without being inundated with formulas. Also, the main statistical platform this text is using in all chapters.

Activity Alert

1   Debate the pros and cons of Quantitative versus Qualitative analyses
2   Create a Quantitative Research Design and identify the IV and DV, as well as any controls
3   Write out research hypotheses and identify the IV and DV
4   Create a nominal, ordinal and interval-ratio variable? Or find a data set in SPSS and identify the variables level of measurement

Test Your Knowledge

1   What is **true** of descriptive statistics?

a   They are a *data reduction* technique
b   They do single variable analyses and simple bivariate testing of relationships, as well as tests for spuriousness
c   They examine the trends and patterns of the data
d   They work with the basic levels of measurement, mainly nominal and ordinal, especially in relationship testing
e   All of the above hold true of descriptive stats

2   What constitutes an inferential statistical test?

a   Zero-order cross tabs
b   Elaborated cross tabs
c   Frequencies
d   t-test, ANOVA, Two-way factorial ANOVA and Regression
e   Univariate statistics

3   The Central Limit Theorem states that if repeated samples of size N are drawn from any population, with mean μ and standard deviation σ, as N becomes larger, the sampling distribution of sample means will approach normality, with mean μ and standard deviation of $\sigma/\sqrt{N}$. Is this statement true or false?

a   True
b   False

4   _____ statistics is a branch of statistics that allows one to generalize from a sample to a population.

a   Descriptive statistics
b   Inferential statistics
c   Descriptive and Inferential statistics
d   EPSeM (Equal Probability Selection Methods or Random Sampling)
e   Central limit theorem

5 In inference base testing, a _____ level determines the amount of error a researcher is willing to make.

   a   Statistical
   b   Confidence
   c   Dummy coded
   d   Type 1
   e   Type 2

6 Levels of measurement – nominal, ordinal, and interval-ratio – are most important because they inform statisticians or quantitative researchers about the statistical analysis they may run.

   a   True
   b   False

7 _____ hypothesis is the researcher's educated guess of a relationship occurring between x and y.

   a   Null hypothesis
   b   Research hypothesis
   c   Alternative hypothesis
   d   Independent
   e   b and c only

8 Quantitative and Qualitative methodologies are vastly different. The key characteristics of a quantitative researcher are the following:

   a   Deductive, macro, numeric, large samples, and generalizability
   b   Inductive, micro, non-numeric, small samples, and no inferences
   c   Deductive, empathy-driven, large samples, and no inferences
   d   Inductive, micro, non-numeric, large samples and unique case by case thematic analysis
   e   None of the Above

9 My survey question is 'Age' and has ranges, like 10–15, 15–20, etc… What type of variable is this classified as?

   a   Nominal categorical
   b   Interval-Ratio
   c   Ordinal categorical
   d   Ordinal discrete
   e   Continuous

10 Quantitative researchers follow the Scientific Method, follow the research cycle, and engage in non-randomized sampling.

   a   True
   b   False

2 Data Manipulation and Sample Modification Techniques

Introduction to Data Modification Techniques

The unprocessed data in statistics are referred to as the raw data. These are the collected, observed, and measured information received through surveys and questionnaires. This can be in person or online. In the initial stages of data analysis, the raw data is uploaded into the SPSS software making it readily available for statistical analysis. As raw data is prepared for analysis, data modification takes place at many stages of the analysis. This is when various adjustments – like data cleansing, data mining and data transformation – take place. Data modification and manipulation enhance data quality, improve data consistency, and reduces data, which in turn enhances interpretability and improves data analysis.

The two most important windows in SPSS that help with organizing raw data are: Variable View and Data View. Variable View allows you to define and label the variable as it appears in the survey. It provides certain descriptions and attributes pertinent to that variable. At the end of it all, the variable will have a name, label, and value. Alternatively, Data View is where the raw data is inputted and can be seen in the form of a data matrix. It is a tedious, yet important part of the statistical pathway process. Here, the reliability of data, as well as suitability, is evaluated. If this phase is done correctly initially, then the interpretability of your statistical tests will increase statistical success. Once the data is cleansed, named, coded, labelled, given a value, and organized in a manageable manner, the analysis of variables can begin using any branch of statistics, but most often it begins with an examination of trends and patterns using univariate statistics. Exploratory data analysis is the key to better understanding any data set(s).

With respect to the statistical analysis process, data modification or data manipulation is when specific changes are made to certain variables in the data set. The data modification stage in statistical analyses is an integral part of data manipulation and analyses. These techniques ensure that data are packaged appropriately to maximize the *best* outcomes of our statistical analyses or results. Most quantitative researchers engage in data cleansing and data mining to gain a better understanding of the variables in a data set. Transforming data is a very normal and common process and practice prior to analyzing data. Quantitative researchers do these types of transformations to combine, collapse, recode, or dummy code data to make the data manageable, workable, and user friendly. These data transformations set the stage for statistical inquiry. All variables being used in analyses are reviewed thoroughly and based on the results of the original variable; various modifications are made to ensure the variable is in a state of acceptable usage for a given statistical test. Any changes to variables take into

DOI: 10.4324/9781003215691-3

consideration things like missing cases, multiple response attributes, the sample or kind of sample you want to discuss or a change in levels of measurement. Once you have your data and are ready to move forward with the analysis you need to make informed decisions about the data or variables, depending on your statistical analysis.

So, why even bother to modify or manipulate data anyway?

- Every statistical test comes with assumptions regarding types and level of measurement of variables, perhaps you want to modify variables to fit the data to the type of test
- Nominal and ordinal data are readily available in survey research, however, are not always considered ideal for inference-based statistics; these tests are most ideal when we have interval-ratio data-so we may modify the data to fit the level of measurement assumption
- We may simply want to increase observed cell counts of a sample for specific response attributes or perhaps the slight difference in response attributes does not really matter; for example, strongly agree, agree, disagree, strongly disagree can easily modify to two response attributes of agree and disagree-the strongly portion may not matter at the end of the day
- We may want to focus on a subset of a sample only further and not the entire sample, like just investigate and analyze women from the data set and not men

The logic and reason for any data modification in statistics must be met with reasonable considerations and simply needs to make sense. Any data modifications that are done, need to be addressed in any paper or publication in such a manner that if someone takes that same data and decides to replicate the procedure, they can successfully do so. Documentation of any data modification is essential. However, before proceeding with any data modification ask yourself the following questions:

- Which variables will you recode and how?
- Which variables will need to be dummy coded?
- Which variables require a reliability analysis? So that scales can be created?
- Is my analysis examining all cases or some?

Data modification, in my years of teaching, is parallel to beautifying a variable to a somewhat ideal state of usage in statistical tests. It's like adding the final touches to the final variable so it can be utilized in the best possible way in any given statistical analysis. In the social sciences, there are many types of informal data modification techniques, like recoding, dummy coding, scaling, splitting files, and select cases. There are also formal methods of data modification known as Exploratory factor analysis (EFA) and Confirmatory factor analysis. This chapter focuses on both informal and formal methods of data modification, with special attention to EFA.

Informal Data Modification Techniques

One of the most common data modification techniques invented and used in the social sciences is *recoding*. This technique is often not discussed too much but is rather very important to the process of statistical analysis. Recoding is a data modification technique that allows us to be creative with our variables and their corresponding response attributes. In statistics, recoding or collapsing of response attributes is done for three main reasons:

1   To *increase cell count* between groups and response attributes within a variable; if the sample size is small then to avoid a small cell count, we collapse categories to gain a decent sample size or may be the slight difference between response attribute labels can be combined. For example, strongly agree, agree, disagree, and strongly disagree can easily transform into 'agree' and 'disagree'. In this way, response attributes are simplified.
2   To create a *dichotomous variable* so that a two-response attribute is created, like [0] no and [1] yes; 0 and 1 represent the absence or presence of a specific attribute or occurrence. This type of variable is used for binary outcomes or specific statistical tests that require a dichotomous variable. It can also be coded as 1 and 2.
3   To create a *dummy coded variable*. Here, assignment of binary values of 0 and 1 allow the variable to be equivalent to an interval-ratio like variable. Coding of data should most always begin with zero. For instance, a simple variable like:

Have you ever been arrested in your lifetime?
[0] no
[1] yes

Are you married?
[0] no
[1] yes

If you begin coding with zero, then it simplifies the process, where you don't have to recode again. For example, if the coding began at [1] and [2], you must recode to dummy code it. But an intelligent researcher will always include and introduce dummy coded coding at the initial stages of survey research. This is a huge time saver for anyone. The advantage of using a binary coding process is that it provides numerical representation.

4   To simply change an interval-ratio variable to a nominal or ordinal variable so that it can be used in a test that requires a nominal or ordinal type data, such as a crosstabulation or elaborated crosstabulation.

In SPSS, recoding occurs under the 'Transform' tab. It is always a good idea to 'recode into a different variable', as you do not want to rewrite the original variable (i.e., never recode into same variable). Know exactly what the objective of recoding is and begin the process. Utilize the traditional method of writing old codes, new codes and how the variable plans to be grouped. Then begin the process. For example, to recode a 'racial background' variable and create a dichotomous or dummy coded variable from three response attributes, the following process should be followed:

Race of respondent original variable: 'Old Values': [1] whites [2] blacks [3] other

Race of respondent recoded variable: 'New Values' in which the old value of 1 becomes the new value of 0 and 2 and 3 old values become 1 as the new value; click the Add button for all transformations:

[0] whites [1] minorities; all other 'old' values outside the 1–3 range become system missing.

In the above example of 'race of respondent', we clearly see how recoding similar response attributes together results in a more compact and simplified version of the variable. We not only recoded the variable to reduce the response attributes, but also dummy coded it, so the variable, if needed, can be categorized as interval-ratio like. One disadvantage of this process is that detail of response attributes is lost, and the variable becomes a more generalized version of itself. So, in the above example, Race shrinks to whites and minorities. Racial groups are all unique. No one group is the same, however lump summing them into 'minorities' is a bit of a stretch. There should always be logic and reason behind the recode and it must be adequately justified for any statistical test. Lastly, any recode that is done must be defined and labelled, tested, and rechecked. Everything should align with the expectations set out. To increase reliability of the outcome of the recode, a simple frequency of the original variable and recoded variable should take place and the two variables should be compared for any errors that may have occurred. If response attributes do not add up, then there are errors and revising the recode may be a good idea.

Formal Data Modification Techniques: Exploratory Factor Analysis, Reliability Analysis and Scaling

Formalized data modification techniques go beyond simple recoding. Exploratory factor analysis (EFA) falls under the more formalized techniques of data modification. Here, the goal is to uncover the underlying relationships seen with multiple measures or variables measuring a single social phenomenon. First, it allows to ensure that the reliability and validity of all participant answers on a survey are accurately done and there are no nonsensical random answers or bad apples in the data. Second, it allows researchers to bring together common multiple measures in various ways, namely through the formal statistical data modification technique of Exploratory factor analysis (EFA), Reliability analysis and standardized scaling. These techniques identify and uncover patterns in the data. In the social sciences we are heavily dependent on the multi-measure approach, especially for quantitative researchers; in surveys we often use multiple measures or variables to assess a particular social phenomenon, like depression, self-esteem, violence against women, criminal behavior, cyberbullying, mental health etc.; the researcher has several questions asked in various way or indicators to assess that social phenomenon (m1, m2, m3, m4, m5, m6, m7). These measures are created to match the operational definition of the measure or variable; however, it is defined. These measures are usually in the form of a Likert scale (ordinal) or may be nominal in nature and simple yes, no response attributes. One requirement of formalized data modification techniques is that all measures must have similar response attributes going in the same direction. For example, all response attributes should be [0] no [1] yes. If any response attribute is [0] yes [1] no, it must be multiple measures in survey research are like the following:

Measure or variable or survey question	Response attribute
Talk to someone when angry	[0] no [1] yes
Drink when angry	[0] no [1] yes
Take drugs when angry	[0] no [1] yes
Listen to music when angry	[0] no [1] yes
Physical fight when angry	[0] no [1] yes
Verbal argument when angry	[0] no [1] yes
Break something when angry	[0] no [1] yes
Exercise when angry	[0] no [1] yes
Ride a bike when angry	[0] no [1] yes
Smoke when angry	[0] no [1] yes
Think about hurting oneself when angry	[0] no [1] yes
Smoke cigarettes when angry	[0] no [1] yes
Drink alcohol when angry	[0] no [1] yes
Walk when angry	[0] no [1] yes
Pray when angry	[0] no [1] yes

All of these anger variables have a common thread of conceptualization and can be analyzed using formal data modification methods. The use of multiple measures in the social sciences is important for two key reasons.

EFA is a *data reduction* technique that allows for multiple measures or variables to be analyzed and build underlying structure from the original pool of variables or measures. The main goal of factor analysis is to get at the underlying structure of data by allowing researchers to combine multiple measures/indicators/questions that measure almost the same social phenomenon. The underlying structure is referred to as a 'latent construct' (i.e., underlying factor).

By combining a larger set of measured variables, a smaller number of latent variables are produced. Hence, the number of original variables is reduced. The function of the 'latent' factor is to explain the correlation that exists among several different measures. In simple terms, there are two fundamental goals of factor analysis:

1   Summarize patterns of correlations among observed variables.
2   To engage in data reduction process to condense many observed variables to a small subset of relatable factors.

EFA is a statistical data modification method that holds no theoretical bearing of the underlying structure of the variables. Often this is known as the 'scientific escape technique'. Here, there is no predetermined theoretical underlying structures, nor is it imposed on the data like with Confirmatory factor analysis (CFA) which is 'planning driven by theory'. EFA has no knowledge of how the selected variables factor out, it is based merely on an educated guess or hypothetical thinking. EFA eventually may develop a theory or social explanation and through proper process and procedures build standardized scales from the latent structures created.

Requirements for EFA

a   Nominal or ordinal data, most often categorical
b   Variables should be conceptually related. They should all measure ONE social phenomenon
c   Response attributes should be directionally similar.
d   Adequate sample size (i.e., at least 50 cases per variable)

The core statistics to report in the EFA are the following:

The first numbers to report are with respect to the sample size. Each variable must have at least 50 cases per variable. Next, the Kaiser-Meyer-Olkin (KMO) and Bartlett's test of sphericity tells us how well correlated the selected multiple measures are and the adequacy of the sample for factor analysis and ranges from 0 to 1. Most often, any high value > 0.60 or 0.70 indicates that the measures are correlated enough to continue with analysis and is considered acceptable. *Eigen Values determine how many latent structures are produced based on the explained* variance accounted for by a factor. Any eigen value that has an explained variance of greater than 1.00 constitutes a meaningful latent structure that is important to further analyze. Anything below 1.00 does not formulate any latent structure on a factor and thus is disregarded. The scree plot or visual image plots the eigen values and visually depicts the number of latent structures produced. This visual representation identifies the points of significant latent structures and not. The point where the scree plot plateaus is when eigen values fall under 1.00. Finally, each factor analysis is rotated to provide an intelligible solution and readable or easy to interpret. Various rotations are plausible: Varimax, Promax, or Oblimin. However, the most robust one is Varimax rotation. The final rotated factor analytical solution provides the factor loadings for each variable. Any variable or measure that has a factor loading greater than 0.40 loads on a specific factor. This is an essential part of interpreting the factor analytical solution as it groups the multiple measures that belong together on a particular latent structure. Once it is determined which measures fall on which latent structure, operationalization of the latent structure can be done. Each latent structure can be adequately labelled to fit the measures or variables. It is important to note that number of latent structures produced equate the number of reliability analyses. For example, if three latent structures are produced from the rotated matrix, then three reliability analyses will follow suit. Whether you use all of them or not is your discretion.

Reliability Analysis

Upon identification of valid underlying latent structures derived from the EFA model, the internal consistency of measures is checked using a Reliability analysis to ensure variables are measuring what they intend to measure. Basically, reliability analysis assesses how well correlated the measures or variables are with each other. The question of relevance is: Are the items from the latent structures measuring a similar phenomenon or construct?

Here, the statistic that informs us about internal consistency of measures is referred to as the Cronbach's alpha and ranges from 0 to 1, where any value less than 0.60 is problematic and unfavorable. This basic statistic evaluates the how well the measures hang well together. Any value greater or equal to 0.60 or 0.70 deems great internal consistency of measures suggesting that the variables are highly correlated, and one may proceed to the next step of building a standardized scale to create interval-like ratio measures.

The 'alpha if item deleted' column also indicates what measures 'if deleted' inflate or deflate the alpha. A low value is indicative of poor measures and internal consistency. Informed decisions are made, should the original Cronbach's alpha value calculate lower than expected. This statistical is crucial prior to scale building to ensure and validate measures. Knowing measures are reliable confirm the process of scale building into interval-ratio measures.

EFA in SPSS

1   Click on Analyze > Dimension Reduction > Factor
2   Move the multiple measures or variables of interest into the right-hand box
3   Click the Descriptives Button and check off KMO and Bartlett's test of Sphericity, Click Continue
4   Click on the Extraction Button, check of Scree Plot and leave all defaults on
5   Click on Rotation Button Choose Varimax
6   Click on Options Button and suppress small coefficients to .40, this means only loadings above .40 will be shown. This simplifies the output and adds ease to interpretability.

Reliability Analysis in SPSS

1   Click on Analyze > Scale > Reliability Analysis
2   Move the multiple measures with factor loadings greater than 0.40 from each latent structure created to items list
3   Click the Statistics Box
4   Check off: Item, Scale, Scale if Item Deleted and click OKs

Please note that a reliability analysis may be run solely on its own with multiple measures. It is not mandatory to run a factor analysis. One can simply test the internal consistency of multiple measures they are interested in summating together for scale construction.

Key Statistics to Report

1   Descriptive statistics: Discuss sample size
2   KMO and Bartlett's Test of Sphericity: Any value 0.60 or 0.70 or higher is deemed sufficient
3   Total variance explained and Eigen values: Any value greater than 1.00 constitutes a latent structure
4   Factor loadings: Any factor loading greater than 0.40 loads on a single factor
5   Reliability analysis: Cronbach's alpha; any value greater or equal to 0.60 indicates good internal consistency

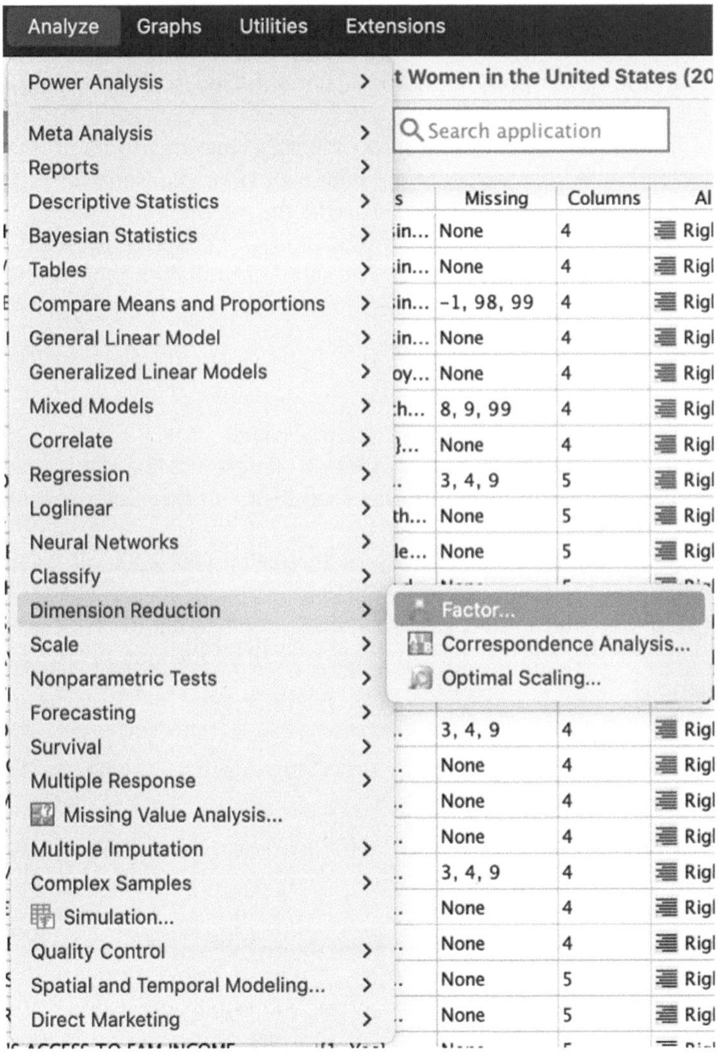

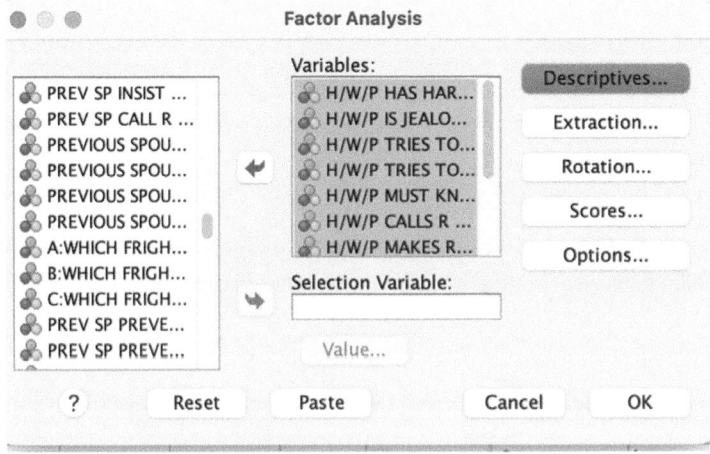

Figure 2.1a SPSS Command of an Exploratory Factor Analysis (EFA)

Figure 2.1b Add Multiple Measure Variables of Interest to the Variables List

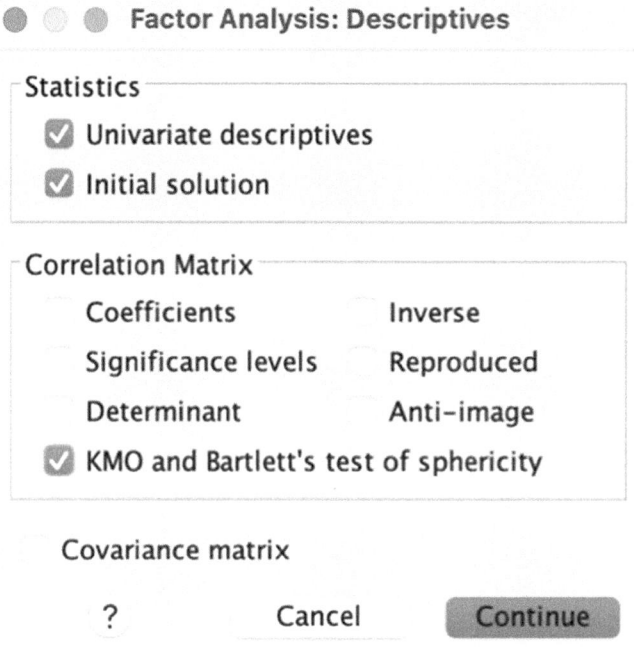

Figure 2.1c Click on Descriptives and Check-off Univariate Descriptives, Initial Solution, and KMO and Bartlett's Test of Sphericity

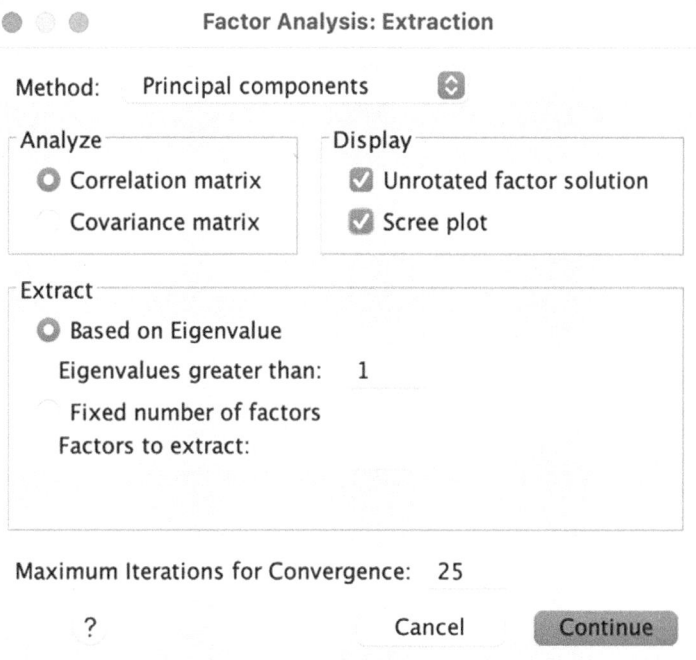

Figure 2.1d Click on Extraction and Check-off Scree Plot and Eigen Values Greater than 1.00

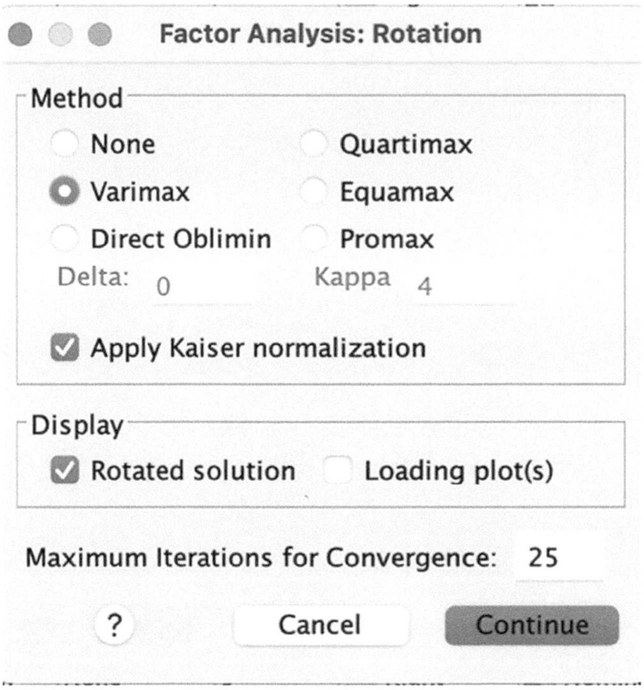

Figure 2.1e Click on Varimax Rotation

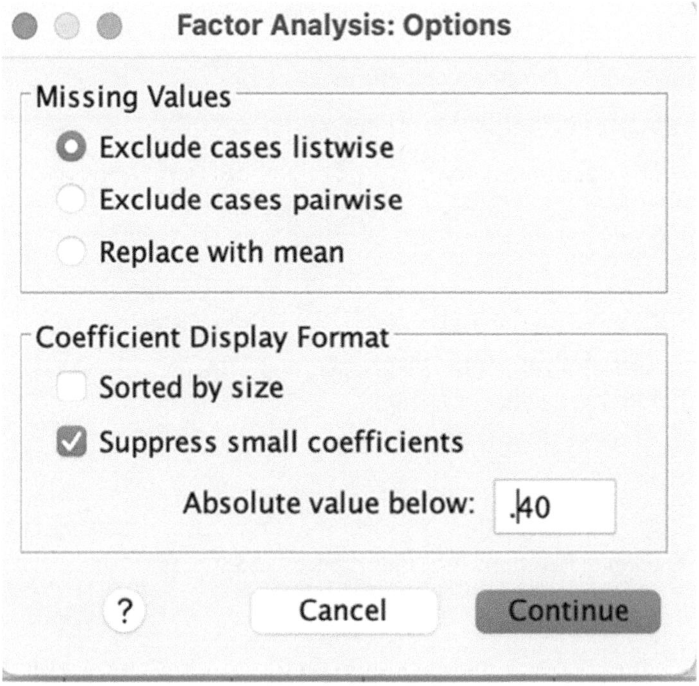

Figure 2.1f Click on Options and Suppress Small Coefficients to an Absolute Value Below 0.40 Factor Loading

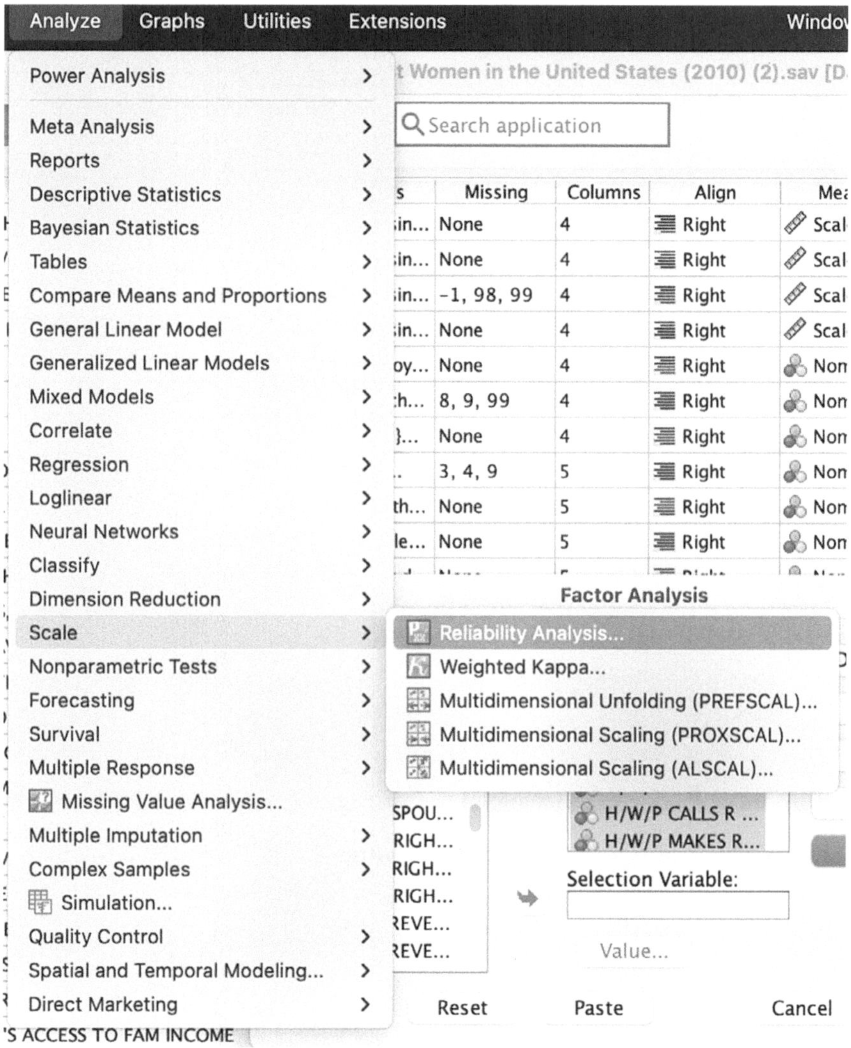

Analyze	Graphs	Utilities	Extensions					Windo

Power Analysis   > t Women in the United States (2010) (2).sav [D.

Meta Analysis   >   Q Search application

Reports   >

		s	Missing	Columns	Align		Me
Descriptive Statistics   > | | in... | None | 4 | Right | Scal |
Bayesian Statistics   > | | in... | None | 4 | Right | Scal |
Tables   > | | in... | None | 4 | Right | Scal |
Compare Means and Proportions   > | | in... | -1, 98, 99 | 4 | Right | Scal |
General Linear Model   > | | in... | None | 4 | Right | Scal |
Generalized Linear Models   > | | oy... | None | 4 | Right | Non |
Mixed Models   > | | h... | 8, 9, 99 | 4 | Right | Non |
Correlate   > | | }... | None | 4 | Right | Non |
Regression   > | | . | 3, 4, 9 | 5 | Right | Non |
Loglinear   > | | th... | None | 5 | Right | Non |
Neural Networks   > | | le... | None | 5 | Right | Non |
Classify   >

Dimension Reduction   >   **Factor Analysis**

Scale   >   🔲 Reliability Analysis...

Nonparametric Tests   >   🔲 Weighted Kappa...

Forecasting   >   🔲 Multidimensional Unfolding (PREFSCAL)...

Survival   >   🔲 Multidimensional Scaling (PROXSCAL)...

Multiple Response   >   🔲 Multidimensional Scaling (ALSCAL)...

🔲 Missing Value Analysis...   SPOU...   🔲 H/W/P CALLS R ...

Multiple Imputation   > RIGH...   🔲 H/W/P MAKES R...

Complex Samples   > RIGH...   Selection Variable:

🔲 Simulation...   RIGH...

Quality Control   > REVE...   Value...

Spatial and Temporal Modeling... > REVE...

Direct Marketing   >   Reset   Paste   Cancel

'S ACCESS TO FAM INCOME

Figure 2.2a Reliability Analysis in SPSS

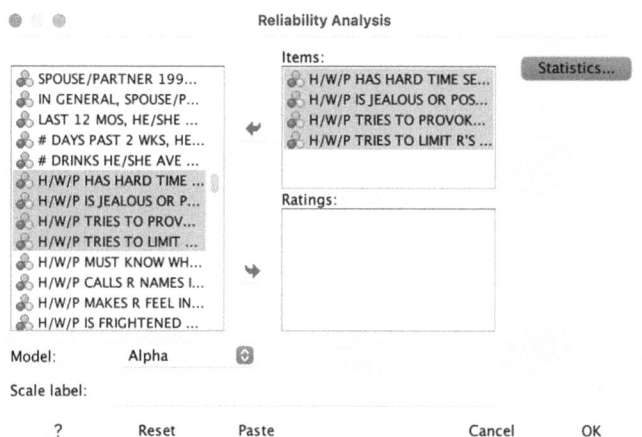

Figure 2.2b Move Multiple Measures to Items List

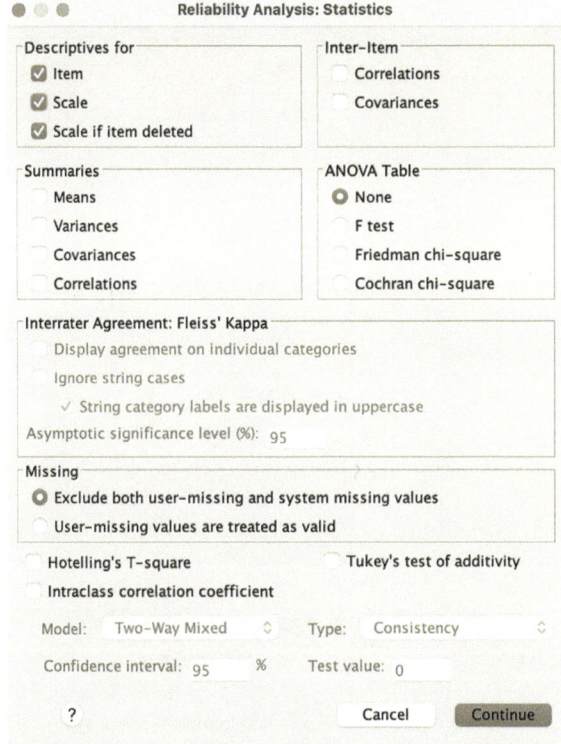

● ○ ○   **Reliability Analysis: Statistics**

Descriptives for
☑ Item
☑ Scale
☑ Scale if item deleted

Inter-Item
☐ Correlations
☐ Covariances

Summaries
☐ Means
☐ Variances
☐ Covariances
☐ Correlations

ANOVA Table
◉ None
○ F test
○ Friedman chi-square
○ Cochran chi-square

Interrater Agreement: Fleiss' Kappa
☐ Display agreement on individual categories
☐ Ignore string cases
   ✓ String category labels are displayed in uppercase
Asymptotic significance level (%): 95

Missing
◉ Exclude both user-missing and system missing values
○ User-missing values are treated as valid

☐ Hotelling's T-square   ☐ Tukey's test of additivity
☐ Intraclass correlation coefficient

Model: Two-Way Mixed ⌄   Type: Consistency ⌄

Confidence interval: 95 %   Test value: 0

?   Cancel   **Continue**

Figure 2.2c Check-off Key Statistics for Reliability Analysis Statistics, Item, Scale, Scale if Item Deleted

Descriptive Statistics

	Mean	Std. Deviation	Analysis N
H/W/P HAS HARD TIME SEE FROM R'S VIEWPT	1.75	.452	5425
H/W/P IS JEALOUS OR POSSESSIVE	1.88	.320	5425
H/W/P TRIES TO PROVOKE ARGUMENTS	1.94	.245	5425
H/W/P TRIES TO LIMIT R'S CONTACTS	1.97	.190	5425
H/W/P MUST KNOW WHO R IS WITH ALL TIMES	1.93	.286	5425
H/W/P CALLS R NAMES IN FRONT OF OTHERS	1.96	.190	5425
H/W/P MAKES R FEEL INADEQUATE	1.95	.258	5425
H/W/P SHOUTS OR SWEARS AT R	1.91	.300	5425
H/W/P FRIGHTENS R	1.98	.139	5425
H/W/P PREVENTS R'S ACCESS TO FAM INCOME	1.98	.166	5425
H/W/P PREVENTS R WORK OUTSIDE THE HOME	1.98	.133	5425

KMO and Bartlett's Test

Kaiser-Meyer-Olkin Measure of Sampling Adequacy.		.858
Bartlett's Test of Sphericity	Approx. Chi-Square	8460.859
	df	55
	Sig.	<.001

Figure SPSS Output #2 Data Modification: Exploratory Factor Analysis and Reliability Analysis

Total Variance Explained

Component	Initial Eigenvalues			Extraction Sums of Squared Loadings			Rotation Sums of Squared Loadings		
	Total	% of Variance	Cumulative %	Total	% of Variance	Cumulative %	Total	% of Variance	Cumulative %
1	3.226	29.330	29.330	3.226	29.330	29.330	2.401	21.828	21.828
2	1.060	9.639	38.969	1.060	9.639	38.969	1.684	15.308	37.136
3	1.028	9.342	48.312	1.028	9.342	48.312	1.229	11.176	48.312
4	.879	7.990	56.302						
5	.840	7.632	63.934						
6	.754	6.855	70.789						
7	.731	6.648	77.437						
8	.692	6.295	83.731						
9	.613	5.573	89.305						
10	.607	5.517	94.822						
11	.570	5.178	100.000						

Extraction Method: Principal Component Analysis.

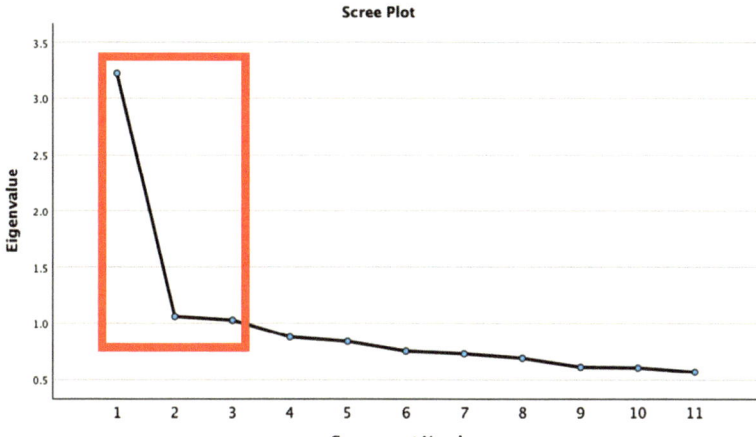

Scree Plot

Rotated Component Matrix[a]

	Component		
	1	2	3
H/W/P HAS HARD TIME SEE FROM R'S VIEWPT	.488		
H/W/P IS JEALOUS OR POSSESSIVE		.764	
H/W/P TRIES TO PROVOKE ARGUMENTS	.649		
H/W/P TRIES TO LIMIT R'S CONTACTS		.484	
H/W/P MUST KNOW WHO R IS WITH ALL TIMES		.787	
H/W/P CALLS R NAMES IN FRONT OF OTHERS	.683		
H/W/P MAKES R FEEL INADEQUATE	.610		
H/W/P SHOUTS OR SWEARS AT R	.677		
H/W/P FRIGHTENS R	.515		
H/W/P PREVENTS R'S ACCESS TO FAM INCOME			.618
H/W/P PREVENTS R WORK OUTSIDE THE HOME			.781

Extraction Method: Principal Component Analysis.
Rotation Method: Varimax with Kaiser Normalization.

a. Rotation converged in 5 iterations.

Case Processing Summary

		N	%
Cases	Valid	5439	68.0
	Excluded[a]	2561	32.0
	Total	8000	100.0

a. Listwise deletion based on all variables in the procedure.

Reliability Statistics

Cronbach's Alpha	N of Items
.662	6

Item–Total Statistics

	Scale Mean if Item Deleted	Scale Variance if Item Deleted	Corrected Item–Total Correlation	Cronbach's Alpha if Item Deleted
H/W/P HAS HARD TIME SEE FROM R'S VIEWPT	9.74	.600	.385	.673
H/W/P TRIES TO PROVOKE ARGUMENTS	9.55	.801	.480	.593
H/W/P CALLS R NAMES IN FRONT OF OTHERS	9.53	.877	.459	.613
H/W/P MAKES R FEEL INADEQUATE	9.54	.819	.403	.616
H/W/P SHOUTS OR SWEARS AT R	9.58	.728	.498	.578
H/W/P FRIGHTENS R	9.51	.962	.335	.649

Reliability

Scale: ALL VARIABLES

Case Processing Summary

		N	%
Cases	Valid	5433	67.9
	Excluded[a]	2567	32.1
	Total	8000	100.0

a. Listwise deletion based on all variables in the procedure.

Reliability Statistics

Cronbach's Alpha	N of Items
.574	3

Item–Total Statistics

	Scale Mean if Item Deleted	Scale Variance if Item Deleted	Corrected Item–Total Correlation	Cronbach's Alpha if Item Deleted
H/W/P IS JEALOUS OR POSSESSIVE	3.89	.157	.416	.437
H/W/P TRIES TO LIMIT R'S CONTACTS	3.81	.260	.330	.564
H/W/P MUST KNOW WHO R IS WITH ALL TIMES	3.85	.175	.445	.371

Technical and Substantive Interpretation of EFA and Reliability Analysis

Exploratory Factor Analysis and Reliability Analysis

Research Purpose: To explore the number of latent structures that are produced from multiple measures extracted from a rotated exploratory factor analytical model and see from the reliability analyses, how many latent structures can be converted to interval-ratio equivalent data and scaled.

Technical Interpretation

Descriptive statistics: The sample size for the multiple measures of husband/wife/partner abusive behaviors is N=5425. These measures work for an exploratory factor analysis solution because they measure a similar social phenomenon, and all response attributes are similar and flow in same direction of 1 being 'yes' and 2 being 'no'. The means and standard deviations make no valuable analytical contributions and therefore are not analyzed.

KMO & Bartlett's Test: The KMO & Bartletts Test indicate that the multiple measures are highly and significantly correlated (KMO = 0.858, $p < 0.05$).

Total variance explained and Scree plot: The total variance explained indicates that there are three 'eigen values' that have variances greater than 1.00 and therefore contribute to the factor analytical solution. There are three ways these measures divide and thus three latent structures are plausible. The scree plot represents the visual image of each eigen value greater or less than 1.00.

(Varimax) Rotated component matrix: The robust intelligible factor analytical solution suggests that any factor loading greater than 0.40 loads on a particular component. For example, hard time seeing from respondent's viewpoint, tries to provoke arguments, calls respondent names in front of others, makes respondent feel inadequate, shouts or swears at respondent, frightens respondent form the first latent structure. The second latent structure comprises of jealous or possessive, tries to limit respondents contacts, and must know who respondent is with all times and the final latent structure accounts for only two measures: prevents respondent access to income and prevents respondent to work outside of home. All three latent structures may be operationalized as the following respectively: toxic partnership, power and control, and financial control.

Reliability analyses: A total of three reliability analyses were run to determine the internal consistency of all latent structures; or how well the measures or variables hang well together? The Cronbach's alpha for all three latent structures were 0.662, 0.574, and 0.999 respectively. Two of the three exceeded 0.60. All but one met the internal consistency. However, the second latent structure could arguably be rounded to 0.60. The Cronbach's alpha if item deleted column did not result in any alpha values higher. Thus, that column was untouched.

Substantive Interpretation

The exploratory factor analytical solution took 11 multiple measures of husband, wife, partner abusive behaviors and engaged in data reduction by creating three latent structures. The factor analytical solution reduced the data to a total of three latent structures from 11 variables. Two of the three latent structures deemed to be fit for scale construction, as the internal consistency of measures were good. Overall, two latent structures were ready to be transformed into interval-ratio like data or its equivalent through standardized scale construction. Univariate statistics for interval-ratio like data will be run on the newly formed variables such that the trends and patterns can be assessed, evaluated and an informed decision can be made.

Unstandardized vs. Standardized Scaling

Unstandardized scaling is simply summating a set of variables together and leaving the scale in its raw form or natural state. Standardized scaling allows you to summate directly observed variables into 'latent' measures, those that are indirectly measured. By taking multiple measures that measure a similar social phenomenon you may summate a set of variables and divide by the number of measures or questions and build a standardized scale, that allows you to compare. There are two core reasons to build scales: 1) To create interval-ratio like data that is continuous in nature from the original nominal or ordinal measures; 2) To summate multiple measures and create one latent construct by dividing by the number of variables or questions used. For example, to build a criminal behavior's scale from four variables or questions in SPSS, the following steps will take place:

1   Locate the set of variables that measure criminal behavior
2   Ensure that all response attributes for each question follows the same direction.
3   Use SPSS to transform the variable into a scale. Transform to Compute; label the variable and summate the variables: $(v1 + v2 + v3 + v4 + v5)/5$; the following command creates a 'standardized' scale, and the response attributes align or match with the original variable scale.
4   You will find this *new* scale at the end of your data set in variable view

If you do not standardize your scale the interpretability of the scale becomes difficult. An unstandardized scale is messy. A standardized scale matches the scale to the original variables response attributes, making the fraction values easier to evaluate. It is always critical to divide by the number of measures to get a lean and readable scale.

Splitting Files and Selecting Cases

Splitting the file and Select Cases in SPSS allows for simple data modifications and further specifies your analysis to a certain group or subset of sample. Splitting the file allows for group comparisons.

Splitting using Compare Groups in SPSS

1   Click Data > Split File.
2   Select the option Compare groups.
3   Double-click the variable Gender to move it to the Groups Based on field.
4   When you are finished, click OK.

Doing the above, allows you to compare groups based on gender response attributes for each question or variable. It provides a more thorough breakdown of numbers by a specific demographic variable. The demographic variable you choose depends on your research objectives. Some common variables used when splitting the file are age, gender, race, education or income.

Select cases is another data manipulation technique that is frequently used in statistics to use a subset of the sample, like men only or women or men who are 18 and older. The 'condition' is based on your research objectives and how you plan to manipulate the sample to achieve your statistical goals. This functionality allows you to include or exclude cases from your analysis. This method becomes useful in research with large data sets because not always are you interested in the entire sample, but a subset of the sample. Sometimes, the sample can be very specific to answer a particular research question.

Select Cases in SPSS

1   Click Data > Select Cases
2   Select 'If condition is satisfied'. Enter a conditional statement, click Continue, and then, OK

When Normality is an Issue and Missing Cases

Normalized data in statistics is a core assumption for many inferential based tests. *Normality* of a distribution is expected in inferential statistics. The shape of your interval-ratio like measures should approach normality and variables should most always be normally distributed. This means that any distribution should produce a bell-shaped curve. Here, the measures of central tendency (i.e., mode, median, and mean) are equal. Any deviation from this results in skewed data or distributions which may impact the outcomes or findings of analysis. These data points are referred to as outliers or extreme scores in a distribution that basically do not fit well with the data. Outliers can be extremely high or low scores and result in skewed data that is positive or negatively skewed. The root cause of outliers in a data set varies. Measurement errors, data entry mistakes, or simple variation in the spread of scores of the data may cause outliers to exist. Sometimes, it becomes essential to normalize the data to the best of our ability using various data transformations for certain statistical tests, like regression analysis. There are three popular methods to do this:

1   Calculate the square root of each datapoint may normalize skewed variables
2   Do a log transformation and calculate the logarithm (Log10/Ln) of each datapoint helps to normalize more strongly skewed data sets
3   Calculating the reciprocal (1/variable) of each datapoint
4   Engage in min-max scaling to transform values of a variable to a specific range
5   Z-score standardization
6   Decimal Scaling

Often data has missing cases for various reasons, and it is not always randomized. The absence of data points for specific variables in a data set can be problematic. It is important to handle these cases carefully and methodically. Not all missing cases need to be removed. However, sometimes certain statistical tests are sensitive to missing cases and need to be eliminated prior to proceeding further with the analysis. Having data that has not dealt with missing data may result in biased or ineffective estimates. This may alter the accuracy of results and may result in Type 1 or Type 2 errors that are misleading conclusions about the data. Document how missing cases were handled. The more transparent you are about it, the better it is for everyone.

Activity Alert

1   Try recoding and dummy coding variables and provide the justification for it.
2   Try to create a codebook and data set in SPSS as a primary investigator.
3   Find a data set in SPSS and run an exploratory factor analysis, reliability analysis and scale a few of the latent structures.

Final Thoughts

Raw data requires refinement and attention to detail prior to basic or advanced analyses. Data modification and manipulation techniques lie at the heart of statistics and can be a book in and of itself. Whether it is doing informal techniques like recoding, dummy coding or scaling or more formal techniques like EFA, splitting and selecting cases, each has its own merit and importance. No one technique is better than the rest. It simply depends on what the statistical test demands and what the research objectives are. Engaging in these data modifications is not enough unless you fully understand the 'why and how' of what you did and knowing if your analyses benefitted from that specific data manipulation or not. Data modifications, most often, begun at the onset of the analysis phase. Also, it is important to note that any data modification done must be discussed in-depth and reported accurately so if any person wants to replicate the process they can easily do so. Modifying samples to fit the analysis is critical to statistics. Moreover, normalizing data and missing cases is just as relevant to thorough statistical analysis, especially as the analysis becomes more inference-based or advanced.

Keywords and Definitions

Recoding	A data modification technique that allows us to be creative with our variables and their corresponding response attributes.
Dummy coding	The process of recoding data into two-response attributes of values 0 and 1, so it displays interval-ratio like characteristics and assumes interval-ratio equivalency.
Dichotomous variable	A two-response attribute variable which can be coded 0 and 1 or 1 and 2.
Exploratory factor analysis (EFA)	A data or dimension reduction technique that allows for multiple measures or variables to be analyzed and build underlying structure from the original pool of variables or measures using a varimax rotation. Often referred to as the 'Scientific Escape' Technique.
Confirmatory factor analysis (CFA)	CFA validates a pre-existing factor structure for a set of variables to test a theory. It works with a pre-determined theoretical structure and is referred to as 'Planning Driven by Theory'. Often, used with Structural Equation Modelling and more advanced techniques.
Reliability analysis	A measure of the internal consistency of variables or how well variables hang together. The sample statistic, Cronbach's alpha should ideally be greater than 0.60.
Standardized scaling	You can summate directly observed variables into 'latent' measures, those that are indirectly measured. By taking multiple measures that measure a similar social phenomenon you may summate a set of variables and divide by the number of measures or questions and build a standardized scale, that allows you to compare.
Splitting files	Splitting the file allows for group comparisons.
Select cases	You can filter out specific individuals and create a subset of the sample and work with those cases only.

| Normality | Normality of a distribution is expected in inferential statistics. The shape of your interval-ratio like measures should approach normality and variables should most always be normally distributed. Any deviation from this results in skewed data or distributions which may impact the outcomes or findings of analysis. |
| Missing cases | When data points are absent in the data set for analysis because of no response, data entry error or problems during data collection phase. Dealing with missing cases in a thorough manner becomes relevant to providing accurate results. |

Test Your Knowledge

After reading this chapter, you should be able to take the test:

1 The _____ data is unprocessed data.

   a   Quantitative data
   b   Qualitative data
   c   Raw data
   d   Descriptive data
   e   SPSS data

2 Exploratory Factor Analysis (EFA) is a _____ technique.

   a   Data or Dimension reduction
   b   Difficult
   c   Quantitative
   d   Inference-based
   e   Sample modification

3 We learned about four types of data modification in this chapter. What are they?

   a   Scaling
   b   Dummy coding
   c   Recoding
   d   Factor analysis
   e   All of the above

4 The factor analysis results in the following:

   a   Manifest structures
   b   Latent structures
   c   Rotated factor analytical solution: VARIMAX
   d   Eigen Values
   e   only b, c and d are correct

5 _____ validates a pre-existing factor structure for a set of variables to test a theory. It works with a pre-determined theoretical structure and is referred to as 'Planning Driven by Theory'. Often, used with Structural Equation Modelling and more advanced techniques.

   a   Exploratory factor analysis
   b   Confirmatory factor analysis

    c   Reliability analysis
    d   Select cases
    e   Standardized scaling

6 _____ measures the internal consistency of variables or how well variables hang together. The sample statistic, Cronbach's alpha should ideally be greater than 0.60.

    a   Statistics
    b   Descriptive statistics
    c   Exploratory factor analysis
    d   Reliability analysis
    e   Dummy coding

7 Standardized scaling allows you to summate directly observed variables into 'latent' measures, those that are indirectly measured. By taking multiple measures that measure a similar social phenomenon you may summate a set of variables and divide by the number of measures or questions and build a standardized scale, that allows you to compare.

    a   a True
    b   b False

8 Which data modification technique allows for group comparisons?

    a   Select cases
    b   Splitting files
    c   Recoding
    d   Dummy coding
    e   Normality

9 _____ of a distribution is expected in inferential statistics. The shape of your interval-ratio like measures should approach normality and variables should most always be normally distributed. Any deviation from this results in skewed data or distributions which may impact the outcomes or findings of analysis.

    a   Select cases
    b   Normality
    c   Standardize scaling
    d   Missing cases
    e   Dummy coding

10 When data points data points are absent in the data set for analysis because of no response, data entry error or problems during data collection phase. These cases are referred to as:

    a   Missing cases
    b   Delete cases
    c   Outliers
    d   Select cases
    e   Recoded cases

3 Describing Trends and Patterns of the Data Using Univariate Statistics

Introduction to Descriptive Statistics Through Univariate Testing

There is an overabundance of information to analyze in the social world and how one chooses to analyze it really depends on the research objectives and research agenda. While there is no universal method to navigate this surplus of information or data, there is a statistical holy grail that most often should be followed to maximize the outcome(s) and, accordingly, make informed decisions. There is a right and wrong way to navigate data sets and the statistical pathways. There needs to be a statistical plan or process in place to explore the statistical madness, and that most often comes from our knowledge pool of survey research methods, statistics and learning how to navigate the basics of data and the analysis that follows. Every day in the world, you use descriptive statistics knowingly and unknowingly. You use these statistics at the grocery store, voting polls, at university when you get your test scores back, discussing the weather or in doing some future financial or market planning. Each day we are engaged in the descriptive world of story-telling through numbers without realizing it. Each task is described numerically through a specific number. That number lies at the heart of descriptive statistics.

In this chapter, we begin to understand and formalize the process of our first instance of data analysis. Now that the data has been mined, cleansed, and transformed, it is ready for the first stage of initial data analysis. Initial data analysis starts with the ABCs of statistics – Univariate Analyses. This chapter highlights the most basic statistical analyses of descriptive statistics utilizing univariate statistics or single variable analysis. Describing the social world through a descriptive statistics lens means that now, the world is being seen as highs and lows or majority and minority or common score or mid-points or average score or spread of the distribution or data points or normal or skewed data. Descriptive Statistics is the preliminary data analysis phase that marks the beginning of the statistical journey. Here, variables and/or survey questions are analyzed through a critical lens. Each outcome is assessed, evaluated, and discussed to make informed decisions about future sophisticated analyses. Even though Descriptive Statistics is not as statistically powerful as other tests, its power lies in the information it conveys and how it conveys it. By reducing data to a single number from large data sets, the basic gist of storytelling begins for each analyzed variable or survey question.

Before understanding exactly what this means, we should begin with the basic definition of what Descriptive Statistics is and how it is defined and understood. In this discussion, you will begin to see how doing statistics works like a process and how certain concepts learned in the earlier chapters come together, especially the notions of levels of measurement and types of variables. You will see that, as you move through the statistical pathways, you bring previous knowledge with you in the process. Hence, it is critical to understand what has been discussed earlier to move on successfully in statistics. This family of statistics, namely, *Descriptive statistics* are a branch of statistics that are engaged in 'data reduction' of collected data. Here, the goal is to take a large data set and reduce it to a single number that explains the nature of the data. It is used to describe the distribution or characteristics or trends and patterns

DOI: 10.4324/9781003215691-4

of a single variable. These statistics try to describe the data or variables by discussing the patterns of data. While they seem very basic, they are key players in the statistical pathway. Every discipline makes use of these statistics. For example, in science it may be used to understand the trends and patterns of trends in medicine or healthcare. In business, assessment of average returns. In political science, these are used to understand the polls. In criminology, patterns of crime can be understood. In psychology, trends in mental health during the COVID-19 pandemic can be assessed using Descriptive statistics. The list goes on and on. Regardless of discipline, these statistics, though not so powerful, provide key information about what's happening with each variable, response attribute and sample size. Through various univariate statistical procedures, it provides the highs and lows of how each participant responded to a specific response attribute. The end goal of descriptive statistics is to present research results in a clear and concise manner by using few numbers, tables, or graphic devices to summarize data. The differences and commonalities seen in the trends and patterns are observed and reported.

As mentioned previously, there are two fundamental branches of Statistics: Descriptive and Inferential statistics. Univariate statistics or single variable analyses falls under Descriptive statistics. These quantitative statistics belong to the family of describing 'trends and patterns' of data through a process of data reduction, one variable at a time. It simply summarizes key features of data by providing summaries about the response attributes and sample. The important thing to remember is that there is no hypothesis testing with an IV or DV here. No relationships or hypotheses are tested here, it is only simple single variable analysis that focuses on one variable at a time. Eventually, our outcomes will allow us to build relationships based on the results.

Univariate statistics comprises frequencies, measures of central tendency, and measures of dispersion or variation or heterogeneity. These statistics are often coupled with the visual representations of bar graphs (with percentages) and histograms. Univariate statistics are presented in this order. As you can see, there are plenty of statistics to choose from. All decisions regarding univariate statistics are levels of measurement based. These levels of measurement – nominal, ordinal, interval-ratio – play a critical role in determining what to run when, and which statistic is chosen. Certain univariate statistics go with specific levels of measurement. I begin the discussion with the simplest, yet essential statistic, frequencies.

Frequencies

The first univariate statistic that is worth mentioning is Frequencies. Initial data analysis will begin with these statistics. They are, let's say, the backbone or infrastructure of our statistical outcomes. Frequencies are your basic univariate statistic that is independent of levels of measurement criteria. This means that frequencies can be run on any level of measurement at any point in time, be it nominal, ordinal, interval, or ratio. There are no limitations. The possibilities are endless. For example, race, satisfaction of university residence, or income levels can all have frequencies run on them. All are different levels of measurement, but each one can produce a frequency table of the counts and corresponding percentages. Frequencies organize and summarize data by displaying how often specific scores were obtained or the number of cases in each category (i.e., response attribute) of a variable. It tallies the number of people in each category and answers the very simple question of how many in each response attribute or category. Hence, it provides the researcher with a summary of the trends and patterns of a particular variable. In frequencies, observed counts and valid percentages are reported to see how many respondents are in each category. Valid percentages exclude missing cases and provides the accurate percentage out of 100 and is standardized and comparable to the other response attributes. The valid percentage represents the standardized number. Observed count is the actual number or occurrences for a particular response attribute. These numbers inform us about the highs and lows of the data and where the majority lie and where the minority of respondents fall into. For example, the frequency distribution of ever a victim of domestic violence indicates that 53 or 30.6% were not survivors of domestic violence, compared to 120 or 69.4% that say they are survivors of domestic violence. From these statistics, one can report that most respondents or participants were survivors of domestic violence, compared to their counterparts. There was a handful that were not. These numbers also provide a sense of how participants answered a particular question and provides insight into the variation of response(s). From this example on domestic violence, there is variation in the data. One thing to remember about running frequencies in SPSS is to organize your data by grouping them by level of measurement, so all nominal variables are run together, followed by ordinal variables and interval-ratio variables. Because every level of measurement is specific about the kind of statistic run, it makes it easier to group variables by level of measurement. This is also a time saver and a systematic method of running these statistics. Begin your analysis with the order of hierarchy of levels of measurement – nominal, ordinal and interval-ratio – and group accordingly. Being organized about how you analyze initial statistical analysis makes a world of a difference, especially when dealing with larger data sets.

Visual Representations: Graphing Frequencies

Frequency distributions are tabled but most always require a visual depiction to complement the table. Charts or graphs are used by researchers to present their data in ways that are visually pleasing, compared to frequency distributions. These graphic displays are quite useful for conveying where the majority or minority of sample lies or an impression of the "overall" shape of a distribution and for highlighting any clustering of cases in a particular range of scores. Although there are a wide variety of graphs (i.e., pie, line charts, stem-leaf plots) two of the most popular ones are: Bar charts with percentages and Histograms with normal curve imposed.

a   *Bar charts* are a graphic display device for nominal/ordinal variables. Categories are represented by bars of equal width, the height of each corresponding to the number (or percentage) of cases in a category.
b   *Histograms* are a graphic display device for interval-ratio variables. Intervals are represented by contiguous bars of equal width, the height of each corresponding to the number (or percentage) of cases in a category.

Together, frequency distributions and graphs summarize the overall shape of a distribution of scores in ways that can be quickly understood by helping us understand the trends and patterns of the variables we are testing. Generating summarized data tables and graphs help make informed decisions about the variables we choose for more sophisticated analyses later. No matter what statistical test you choose to run, basic frequencies should be part of the first analysis. Once frequencies are analyzed, the next stage of univariate statistics are measures of central tendency which provide a unique perspective that is indeed worth paying attention to and reporting when analyzing data.

Measure of Central Tendencies: The 3 Ms

Measures of central tendency (MCT), which are levels of measurement dependent provide researchers with some idea of the typical, common, mid-point or average case in a distribution. The three measures of central tendency are: *mode, median*, and *mean*. Measures of central tendency do not provide information on variability of data or spread of scores. These are level of measurement specific. All are powerful, as they can reduce huge arrays of data to a single, easily understood number.

a. *Mode* is the most common or frequently occurring number in a distribution. It is used

with nominal categorical data. Here, you discover what the modal category or the most common score of any distribution is. In a frequency distribution, the mode is easy to locate. To calculate mode, arrange data in ascending order and count how many times each score occurs. Data sets can be unimodal (one mode or common score) or bimodal (two modes or two common scores). For example, if 60% of men smoke and 40% of women smoke, the modal category for this variable is men. Men are the most frequently occurring response or common score. Women do not represent the modal category because they are not the most common score in the distribution.

Another example of a unimodal distribution: 1, 2, 3, 4, 4, 4, 4, 5, 6, 7, 8, 9, 10. Here 4 is the modal category to report.

An example of a bimodal distribution: 1, 1, 1, 2, 3, 4, 4, 4, 4, 5, 6, 7, 8, 9, 10. Here 7 and 9 are the modal categories to report.

Use the Mode when:

- Variables are measured at the nominal level.

b. *Median* is the mid-point or 50th percentile (when examined as quartiles) of a distribution and is often used with ordinal and interval-ratio data. Here, the goal is to report the central score of a distribution. Thus, if the median family income for a community is $50,000, half the families earn more than $50,000 and half earn less. The median is not sensitive to extreme scores. To find the median (by hand), the cases must be placed in *order* from the highest to lowest score or vice versa. Use the formula: N + 1/2 10, 20, 30, 40, 50 (5 data points arranged in ascending order); 5 + ½ = 6/2 = 3rd score

This indicates that the 3rd score is the position of the median value. In this case, the value is 30.

Use the Median when:

- Variables are measured at the ordinal level categorical or discrete data
- Variables measured at the interval-ratio level and have highly skewed distributions
- You want to report the central score of a distribution

c. *Mean* simply put, is the arithmetic average of a distribution and is sensitive to outliers or extreme scores. It pulls in high and low directions of extreme scores. The mean is the most informative measure of central tendency and work with the highest level of measurement, interval-ratio data, or its equivalent.

$$\bar{X} = \frac{\sum X}{N}$$

For example: 10, 15, 20, 25, 30, 35, 40
10 + 15 + 20 + 25 + 30 + 35 + 40/7 = 25

Use the Mean when:

- Variables are interval-ratio continuous data

Normal Curve Properties or Skewed Data

The normal curve is a theoretical, bell-shaped, symmetrical curve that is at the crux of statistical analysis. In real life, this type of curve is somewhat superficial and relatively non-existent. Distributions can approximate the normal curve but not be completely identical to it or be a mirror image of it. It is important to note how the normal curve and mode, mean and median values play a role. The normal curve is a product of the mode, median, and mean being all the same. When they are all equal the histogram distribution assumes normal characteristics, like it is bell-shaped, unimodal, symmetrical with average tails and there is no skewed data or outliers or extreme scores. However, when there is a shift in these values, specifically the median and mean values, there is skewness of the distribution in a positive or negative direction.

The normal curve can have extreme scores that shift to the left or right. Skewness is a lack of symmetry or lopsidedness of a distribution. In statistics, a positive skew occurs when a distribution has some extremely high scores or longer right tail than left. Here, the mean will always have a greater numerical value than the median.

Example: May occur when you have a test that is very difficult, and few people get scores that very high and many more get scores on the low end. Here, the mean will be lower in value than the median.

Example: It may occur when an easy test is given … lots of high scores and few low scores.

In the case of positive skewness that occurs when there are extremely high scores or a longer right tail than left. Here, the mean will always have a greater numerical value than the median. We could see this when you write a difficult test and only few students achieve a high score. In the case of negative skewness, the exact opposite occurs. Here, the distribution has some extremely low scores or a longer left tail than right. In this scenario, the mean will be lower in value than the median value. Again, this may occur when an easy test is given and there are lots of high score and few lower scores.

Positive skewness rule	Negative skewness rule
If the mean > median, the distribution is positively skewed.	If the median>mean, the distribution is negatively skewed.

Again, measures of central tendency provide and summarize data by focusing on the common score, midpoint, and average, without any sense of variability or spread of scores. Measures of dispersion aid in assessing the spread of scores.

Measures of Dispersion

Measures of dispersion, variation or heterogeneity assess variability of the data or the spread of scores for only interval-ratio continuous data or its equivalent. This is relatively different from measures of central tendency. Again, there are various ways that one can assess variability of data in distributions, however the common approach is to assess variance and its corresponding standard deviation. Standard deviation is the product of the square root of variance and most often entails a series of calculations to get.

The range is one of the basic measures of dispersion that assesses spread of scores by taking the difference of the minimum and maximum values in a data set. It simply provides a range of the highest value and lowest value in a distribution. This provides a sense of variability of a distribution but is not a very powerful statistic.

Variance (s^2) is the most frequently used measure of variability and is defined as the mean of the squares of deviation scores. Overall, the variance is a kind of average of how much scores deviate from the mean after they are squared. Variance provides a measure of the variability. The more variability in a group, the higher the value of the variance; the more homogeneous the group, the lower the variance. Standard deviation (s) is simply the square root of the variance or of the squared deviations of the scores around the mean, divided by N of a sample. In simple terms, It's the average amount of variability in a set of scores or the average distance from the mean. For example, if the variance is 25, SD = 5. The values of "s" indicate relative variability within a group. The larger the standard deviation value the greater the heterogeneity or variability of data in a distribution. The notion of variability is what makes a distribution exciting to look at. Knowing that there is variation in the data is good. It provides an indicator that not everyone answered the question in a similar manner and the data is indeed interesting. If all participants answer in a similar way, the data become uninteresting and dull. The variation in data is what makes statistical outcomes interesting.

There are many ways to describe the variation of data at the univariate level. However, most often it is discussed in terms of kurtosis, how flat or peaked a distribution appears. While there are many forms of kurtosis the most common used to describe histograms for interval-ratio data or its equivalent are:

1   Platykurtic in which the distribution is relatively flat compared to a normal distribution and thus displays increased variation in scores.
2   Leptokurtic is a distribution that is relatively peaked compared to a normal distribution and thus display decreased variation in scores.

Summary of Univariate Descriptive Statistics Reporting by Levels of Measurement

Lowest level of measurement	*Level of measurement*	*Statistic to report*	*Graph*
	Nominal (categorical)	Frequencies Measure of Central Tendency: Mode	Bar chart with percentages
	Ordinal (categorical or discrete)	Frequencies Measure of Central Tendency: Median	Bar chart with percentages
	Interval (continuous)	Frequencies Measure of Central Tendency: Median and Mean Measure of Variation: Range, Variance, Standard Deviation Quartiles Skewness (+ve/-ve Skewness) and Kurtosis (flat/peaked)	Histogram with normal curve
Best level of measurement	*Ratio (continuous)*	Frequencies Measure of Central Tendency: Median and Mean Measure of Variation: Range, Variance, Standard Deviation Quartiles Skewness (+ve/-ve Skewness) and Kurtosis (flat/peaked)	Histogram with normal curve

Activity Alert

1   Open any data set in SPSS and identify the levels of measurement and types of variables.
2   Write down the rules of running Univariate Statistics.
3   What is the difference between skewness and kurtosis.
4   Run Univariate Statistics in SPSS using any data set. Interpret the Results.

Univariate Statistics in SPSS

1   Click on Analyze > Descriptive Statistics > Frequencies
2   Move the variable of interest into the right-hand box
3   Click Statistics, select Measures of Central Tendency and Measures of Dispersion
4   Click on the Chart button, select Bar Chart with Percentages or Histograms with Normal Curve, and press the Continue button.
5   Click OK to generate a frequency distribution

Key Statistics to Report

1   Sample size: Valid and missing cases; discuss all valid and missing cases in the distribution
2   Frequency counts and Valid percentages: Provide all frequency counts and valid percentages for each response attribute or simply provide trends and patterns, highs or lows, or majority or minority cases
3   Measures of central tendency: Discuss mode, median and mean values by level of measurement to provide a sense of typical score, mid-point, and average
4   Measures of dispersion: Discuss Standard deviation only for interval-ratio data and provide a discussion regarding variation of the distribution or spread of scores.
5   Refer to Graphs: Bar or Histogram

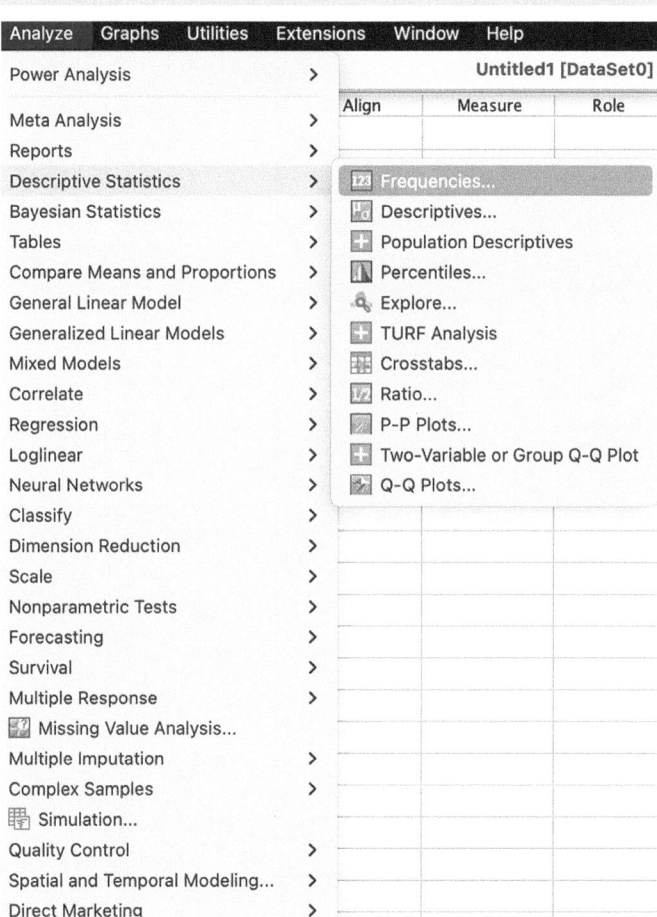

Figure 3.1a Univariate Statistics in SPSS

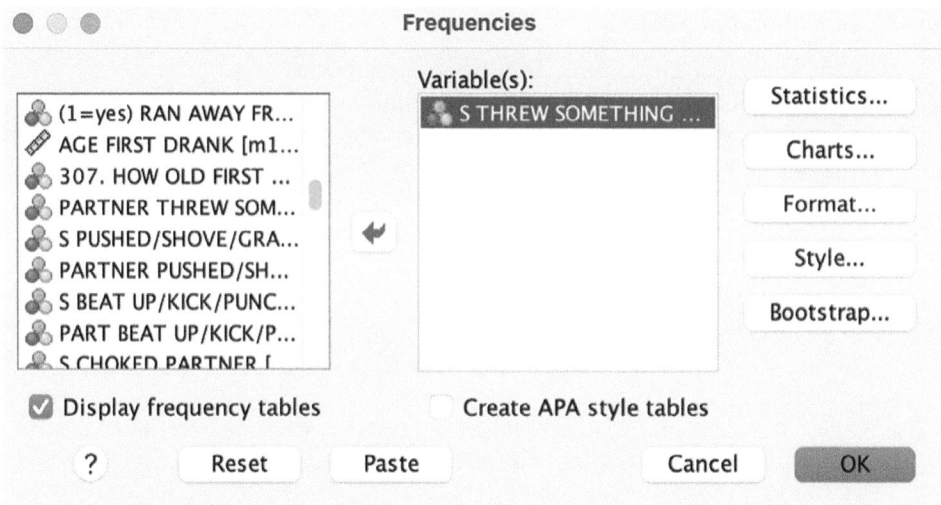

Figure 3.1b Move Variable(s) of Interest by Levels of Measurement into Right Hand Box and Click Statistics

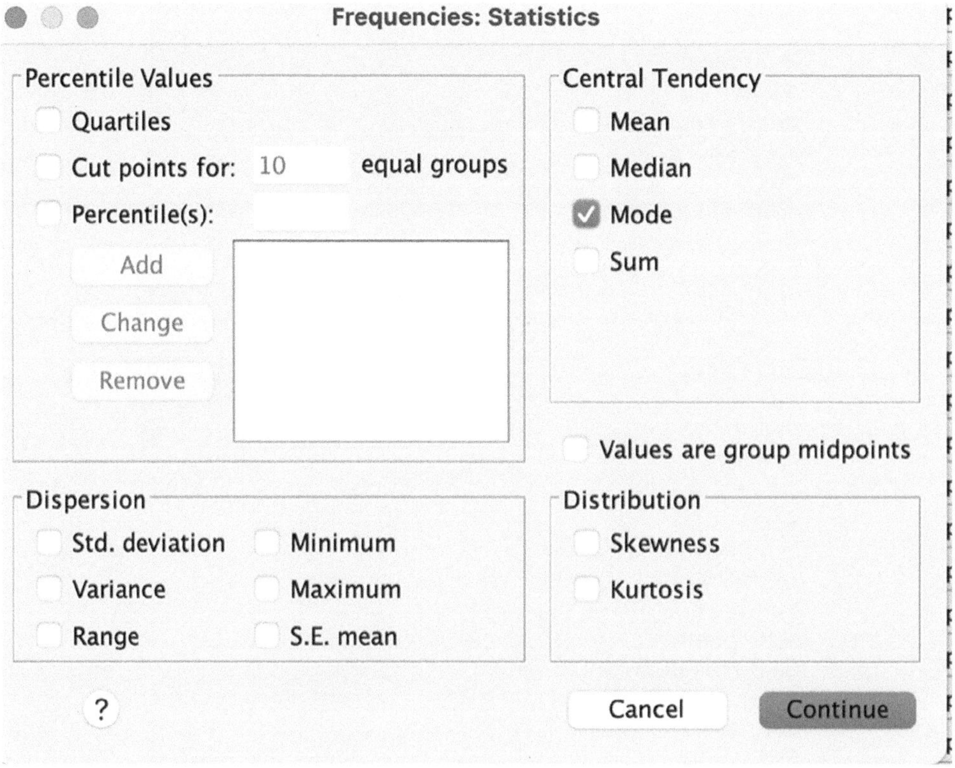

Figure 3.1c Check-off Key Univariate Statistics, Measures of Central Tendency and Measures of Dispersion or Variation by Levels of Measurement

Frequencies: Charts

Chart Type
- None
- ● Bar charts
- Pie charts
- Histograms:
  - Show normal curve on histogram

Chart Values
- Frequencies   ● Percentages

?   Cancel   Continue

Figure 3.1d Check-off Visual Representation by Selecting Bar Chart with Percentages or Histogram with Normal Curve in Accordance with Level of Measurement

Statistics		
NOMINAL-CATEGORICAL: S THREW SOMETHING AT PARTNER THAT HURT		
N	Valid	173
	Missing	1
Mode		0

Figure SPSS Output #3 Example of Univariate Statistics

		Frequency	Percent	Valid Percent	Cumulative Percent
Valid	No	99	56.9	57.2	57.2
	Yes	74	42.5	42.8	100.0
	Total	173	99.4	100.0	
Missing	Missing	1	.6		
Total		174	100.0		

S THREW SOMETHING AT PARTNER THAT HURT

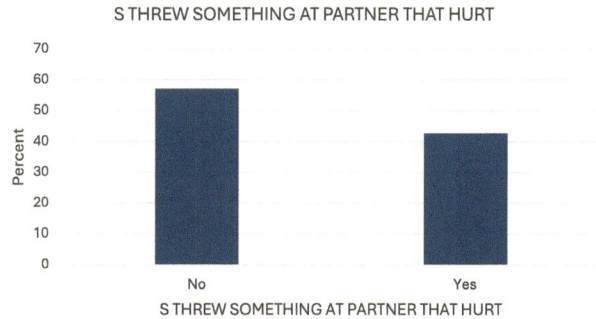

S THREW SOMETHING AT PARTNER THAT HURT

S THREW SOMETHING AT PARTNER THAT HURT

ORDINAL-CATEGORICAL: OFTEN S USED FORCE TO DEFEND HERSELF			
N	Valid		117
	Missing		57
Median			1.00

OFTEN S USED FORCE TO DEFEND HERSELF					
		Frequency	Percent	Valid Percent	Cumulative Percent
Valid	Never	25	14.4	21.4	21.4
	Occasionally	45	25.9	38.5	59.8
	Most of the time	16	9.2	13.7	73.5
	All of the time	31	17.8	26.5	100.0
	Total	117	67.2	100.0	
Missing	Missing	1	.6		
	NA	56	32.2		
	Total	57	32.8		
Total		174	100.0		

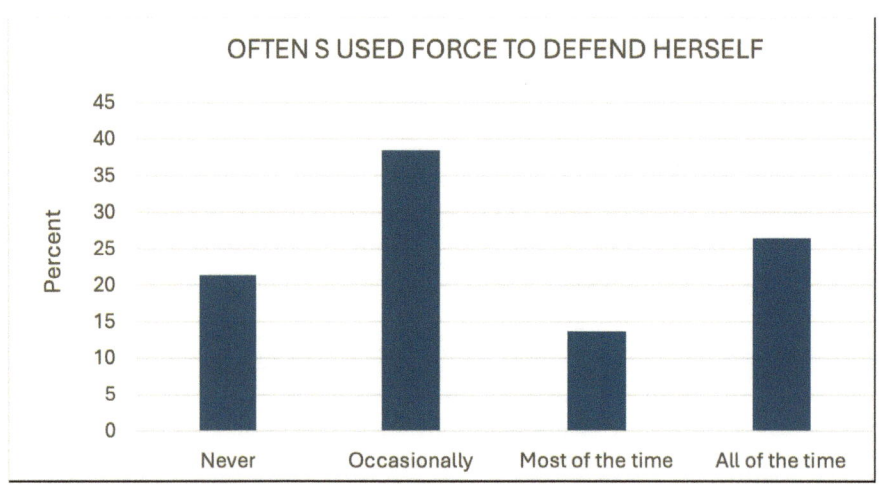

OFTEN S USED FORCE TO DEFEND HERSELF

How Old First Time Had Boyfriend

INTERVAL-RATIO CONTINUOUS: HOW OLD FIRST TIME HAD A BOYFRIEND			
N	Valid		173
	Missing		1
Mean			14.71
Median			15.00
Std. Deviation			2.358
Variance			5.558
Percentiles	25		13.00
	50		15.00
	75		16.00

307. HOW OLD FIRST TIME HAD A BOYFRIEND		Frequency	Percent	Valid Percent	Cumulative Percent
Valid	5	1	.6	.6	.6
	9	1	.6	.6	1.2
	10	1	.6	.6	1.7
	11	6	3.4	3.5	5.2
	12	20	11.5	11.6	16.8
	13	27	15.5	15.6	32.4
	14	17	9.8	9.8	42.2
	15	38	21.8	22.0	64.2
	16	35	20.1	20.2	84.4
	17	9	5.2	5.2	89.6
	18	9	5.2	5.2	94.8
	19	6	3.4	3.5	98.3
	20	1	.6	.6	98.8
	21	1	.6	.6	99.4
	25	1	.6	.6	100.0
	Total	173	99.4	100.0	
Missing	NA	1	.6		
Total		174	100.0		

Histogram

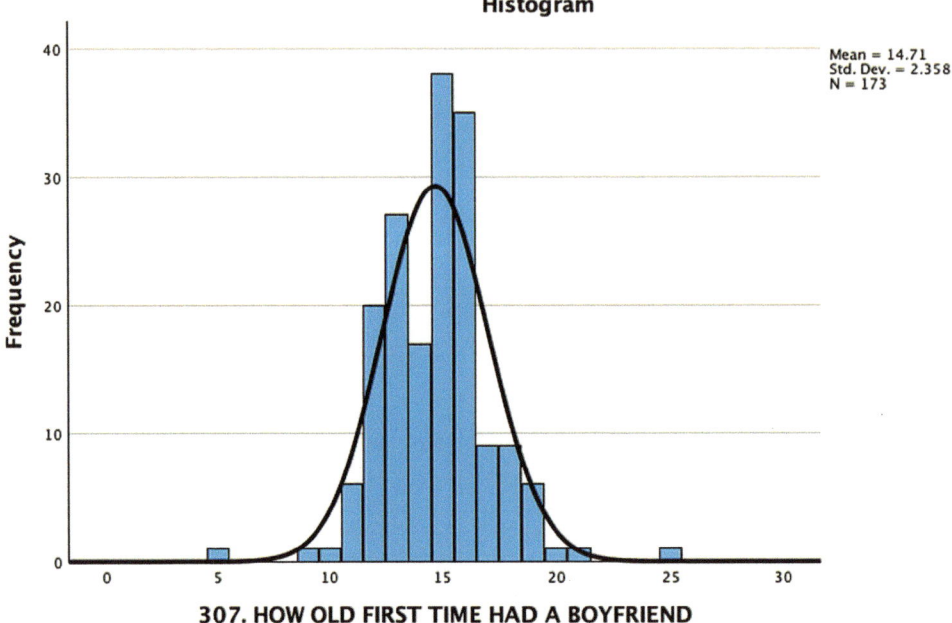

307. HOW OLD FIRST TIME HAD A BOYFRIEND

Technical and Substantive Interpretation for Univariate Analyses

Research Purposes: *To examine the trends and patterns of the data for the nominal, ordinal, and interval-ratio variable.*

Nominal Categorical Variable: S Threw Something at Partner that Hurt

Technical: The total sample size is 173 and there is one missing case. The modal category is 0, which is 'no'.

The frequency distribution of the nominal categorical variable, 'subject threw something at partner that hurt', indicates that 99 (57.2%) of respondents did not throw something at their partner that hurt them. Only 74 (42.8%) reported that they did throw something at partner that hurt. The bar chart (with percentages) depicts the visual representation for this frequency distribution (see SPSS #3 Output 3a–3c).

Substantive: Therefore, most respondents reported not throwing something at their partner that hurt them. A handful of participants stated they did indeed throw something at their partner that hurt them. The most common response attribute is "no" and the subject did not throw something at partner than hurt. Thus, based on the data the trends and patterns indicate that most participants were not violent.

Ordinal Categorical Variable: S Often Used Force to Defend Herself

Technical: Approximately 117 (N) respondents answered this question, with 57 missing cases. The median category or midpoint or 50th percentile of the distribution is 1, which is "occasionally". The frequency distribution of the ordinal categorical variable, often subject used force to defend herself denotes that 25 (21.4%) never used force to defend themselves, whereas 45 (38.5%) occasionally did use force to defend themselves. However, 16 (13.7%) or 31 (26.5%) spoke of defending themselves using force most and all the time, respectively. The bar chart (with percentages) depicts the visual representation for this frequency distribution (see SPSS Output #3 3d–3f).

Substantive: Consequently, the frequency distribution shows much variation in the use of force to defend themselves. For example, most respondents reported "occasionally" defending themselves using force, followed by those that used force "all the time" to "never" and "most of the time". The trends and patterns of this frequency distribution indicates a varied response in their use of force to defend themselves. Overall, 50% of respondents fall above "occasionally" and state "never" and 50% fall below, stating "most or all of the time".

Interval-Ratio Continuous Variable: How old were you the first time you had a boyfriend

Technical: The frequency distribution of the interval-ratio continuous variable, how old were you when you first had a boyfriend, indicates that 38 (22%) of women in the sample had their first boyfriend at 15 years of age, followed by 16 (20.2%) of women. There were some in the sample that claimed to have a first boyfriend at 12 (20 or 11.6%), 13 (27 or 15.6%), 14 (17 or approximately 10%) respectively. There were very little women who claimed to have their first boyfriend at 5, 9 and 10 years old respectively (1 or 0.6%). There were about 5.2% (9) 17- and 18-year-olds that claimed their first relationship. Few women reported having their first boyfriend at 20, 21 and 25 (1 or 0.6%). The histogram with normal curve depicts the visual representation for this frequency distribution (see SPSS Output #3 3i). The distribution shows a slight negatively skewed, with extremely low scores, with the median (15.00) being greater than the mean (14.71) value. The distribution is "peaked" and reflects a leptokurtic curve with minimal variation (s = 2.358 years) (see SPSS Output #3 3g–3i).

Substantive: Consequently, the frequency distribution shows slight variation in the use of force to defend themselves. For example, most respondents reported "occasionally" defending themselves using force, followed by those that used force "all the time" to "never" and "most of the time". The trends and patterns of this frequency distribution indicates a minimally varied response in their use of force to defend themselves. There are extremely low ages for first time had boyfriend.

Final Thoughts

Univariate statistics describe the social world around us and provide us with many statistical tests and options to do this. These statistics are descriptive statistics that simply provide descriptions of single variables, like gender, race, years of education, grade point average, the # of times you have been arrested, etc. These statistics *describe* the trends and patterns of data using very basic statistical techniques, like frequencies, graphs, measures of central tendency and measures of dispersion/variation/heterogeneity. While some of these statistics are levels of measurement dependent, they should always be run in an *orderly* manner to organize and enhance the interpretability of data. Understandably, no relationships or hypotheses are tested for. There are no independent or dependent variables being tested simultaneously. Simple statements regarding distributions in the data are accounted for and the goal is to parsimoniously describe each variable on its own terms in accordance with level of measurement rules. The goal at this stage is two-fold: to explore the trends and patterns of the data for each variable and to try to establish a justification for future sophisticated analyses. The goal here is not generalizability or making inferences of any kind. It is basically taking the data for what it is and engaging in data modifications or manipulations before getting to the next stage. Also, ensuring that the sample size is adequate to progress to the next stage and that missing cases are handled with care. Any univariate statistics run, must keep in mind levels of measurement of all variables to ensure correct statistics are checked off for each. In a nutshell, univariate stats are the heart of any analyses. These are a must to get a sense of the data before moving to more advanced statistical techniques. It really gives face to the raw data at hand and allows for the beginnings a quantitative story telling session. One must always remember that these statistics must always be run, prior to any other sophisticated or advanced analysis.

Keywords and Definitions

Descriptive statistics	A branch of statistics that are engaged in "data reduction" of collected data. The goal is to take a large data set and reduce it to a single number that explains the nature of the data. It is used to describe the distribution or characteristics or trends and patterns of a single variable.
Univariate statistics	Single variable analyses using statistical tests like frequencies, measures of central tendency, and measures of dispersion or variation or heterogeneity. These statistics are often coupled with visual representations of bar graphs (with percentages) and histograms.
Levels of measurement	Nominal, ordinal, interval-ratio, play a critical role in univariate statistics in informing about the statistics that is suitable for a particular variable.
Frequencies	The heart of univariate statistics. They organize and summarize data by displaying how often specific scores were obtained or the number of cases in each category (i.e., response attribute) of a variable by providing information on the observed counts and valid percentages for each category.
Measures of central tendency	These describe the common score, mid-point, or average score in a distribution. They do not discuss spread of scores in a distribution.
Mode	The most common score in a distribution and best used with nominal categorical data.
Median	The middle point or 50th percentile of a distribution and best used with ordinal and interval-ratio data.
Mean	The arithmetic average of a distribution which is used with the highest level of measurement, interval-ratio data. Means are sensitive to outliers or extreme scores.
Normal curve	A product of the mode, median, and mean being all the same. When they are all equal the histogram distribution assumes normal characteristics, like it is bell-shaped, unimodal, symmetrical with average tails and there is no skewed data or outliers or extreme scores.
Skewness	A lack of symmetry. Data may be positively skewed or negatively skewed.
Positive skew	The data set has extremely high scores.
Negative skew	The data set has extremely low scores.

Measures of variation or dispersion or heterogeneity	These describe the spread of scores in a distribution for interval-ratio continuous data.
Range	The most basic measure of dispersion and report the minimum and maximum values of any data set.
Variance	(s2) is the most frequently used measure of variability and is defined as the mean of the squares of deviation scores. Overall, the variance is a kind of average of how much scores deviate from the mean after they are squared.
Standard deviation	(s) is simply the square root of the variance or of the squared deviations of the scores around the mean, divided by N of a sample.
Platykurtic	The distribution is relatively flat compared to a normal distribution and thus displays increased variation in scores. There is increased spread or variation in these distributions.
Leptokurtic	The distribution is relatively peaked compared to a normal distribution with less variation. There are increased instances of clustering of data, thus less spread of the scores.

Test Your Knowledge

1   Univariate Statistics are a _____ technique.

   a   Standard
   b   Data reduction
   c   Relationship building
   d   Bivariate
   e   Average

2   Sometimes, data may be skewed and we need it for the analysis. Skewed data may negatively influence our outcomes, especially for inferential statistics. What variable transformations can be done to overcome this hurdle?

   a   Replace the variable
   b   Calculate the square root of each data point
   c   Take the 'log10' of each variable
   d   Calculate the reciprocal of each data point.
   e   b, c, and d only

3   What is **true** of descriptive statistics?

   a   They are a data reduction technique
   b   They do single variable analyses and simple bivariate testing of relationships, as well as tests for spuriousness
   c   They examine the trends and patterns of the data
   d   They work with the basic levels of measurement, mainly nominal and ordinal, especially in relationship testing
   e   All of the above hold true of descriptive stats

4   An instructor is preparing a report showing mid-semester grades. They note that the mean, median, and mode are all exactly 55.00. What may she conclude about the data?

a   The distribution of grades is unskewed (there is no lack of symmetry)
b   There is a negative skew in the distribution
c   There is a positive skew in the distribution
d   There is a leptokurtic and platykurtic kurtosis in the distribution
e   a and d

5   _____ measures spread of a distribution.

a   Skewness
b   Kurtosis
c   Measures of central tendency
d   Measures of variation
e   Univariate statistics

6   What is a limitation of univariate descriptive statistics?

a   It does not test relationships
b   It does *not* allow us to generalize from a sample to a population
c   There are no visual representations of it, like bar charts or histograms
d   a and b
e   a and c

7   What does the frequency Table 3.1 suggest?

Statistics

HOMOSEXUAL SEX RELATIONS

N	Valid	885
	Missing	615
Median		1.00

HOMOSEXUAL SEX RELATIONS

		Frequency	Percent	Valid Percent	Cumulative Percent
Valid	ALWAYS WRONG	526	35.1	59.4	59.4
	ALMST ALWAYS WRG	39	2.6	4.4	63.8
	SOMETIMES WRONG	67	4.5	7.6	71.4
	NOT WRONG AT ALL	253	16.9	28.6	100.0
	Total	885	59.0	100.0	
Missing	NAP	519	34.6		
	DK	88	5.9		
	NA	8	.5		
	Total	615	41.0		
Total		1500	100.0		

a   Participants mostly viewed homosexual relations as being "always wrong"
b   About 30% of participants viewed it is a "not wrong at all"
c   From 1,500, 885 responded and 615 cases were missing
d   The median was 1.00 (always wrong)
e   All of the above hold true

8   In Statistics, we discuss univariate histograms with respect their spread of the distribution, especially with interval-ratio data. _____ is the name used to refer to the distribution as peaked or flat.

   a   Skewness
   b   Kurtosis: Platykurtic/Leptokurtic
   c   Mean
   d   Standard deviation
   e   Normal curve properties

9   You have an ordinal variable, what are the appropriate univariate statistics to run?

   a   Mode and bar chart with percentages
   b   Mean and histogram
   c   Frequencies, median, mean, and bar chart with percentages
   d   Frequencies, mode, and histogram
   e   None of the above

4 Bivariate and Multivariate Descriptive Statistics Using Non-Parametric Tests

In the preceding chapter, we learned about descriptive statistics for a single (uni-variate) variable and how important these simple statistics become for analyzing data to make informed decisions about more sophisticated analyses, through exploring the trends and patterns of the data. However, such descriptions only provide us with limited information about sole data distributions, typical averages, and variation or spread in the distribution. These statistics in no way allow us to assess relationships or correlations or influences of other unseen third variables, like control variables. They do not inform of us how two variables vary. Sometimes, our scope about the social world expands beyond one variable analysis when investigation occurs and curiosity sparks. For example, we may be interested in social media usage by gender; or academic success by race; or how years of education influence type of employment and the list goes on and on. What do you think of these bivariate relationship possibilities?

- Length of prison term and satisfaction with the criminal justice system
- Do living arrangements (on or off campus) impact criminal behavior?
- Is there a relationship between prisoners contracting COVID-19 and visitation hours?
- Does your social class influence jail time or sentence length?

Activity Alert

Look around the social world and think of some bivariate relationships you would want to analyze. Also, try and see if you can identify the IV and DV in each relationship. Keep in mind levels of measurement.

Every discipline, whether sociology, psychology, business, criminology, epidemiology, marketing, and science have these types of questions. They are not just specific to the social sciences. When there is a vested interest in social inquiry, instinctively one thinks in terms of bivariate relationships or how to variables come together to establish a relationship of sorts. Bivariate descriptive statistics focuses on the cross-classification of two variables that are nominal or ordinal in nature. Much social research in the social sciences is more concerned in *explaining* than simple describing of variables or a particular distribution. As social scientists we are always interested in going beyond simple data distributions and exploring various research questions or hypotheses in relation to an independent and dependent variable, or sometimes even adding a control variable(s).

DOI: 10.4324/9781003215691-5

Bivariate descriptive associations hold a special place in statistics. Remember, that to be classified as a bivariate association that establishes cause and effect there needs to be three components:

1   Must have temporal order amongst the IV and DV
2   Association between the IV and DV
3   Ensure that no unseen third variable is influencing the relationship (i.e., check for spuriousness)

The focus of this chapter is to discuss the non-parametric test, or what is referred to as distribution free tests, of (Zero-order) Bivariate Crosstabulations at length and to understand the interpretations that surround them. In statistics, there are parametric and non-parametric tests. Parametric tests are usually associated with population parameters. These tests are most often inference-based, and their main objective is to infer findings from a sample to a population parameter, based on a normally distributed population distribution, interval-ratio continuous data and large sample size. Non-parametric tests are not bound to the normal distribution because they work with nominal and ordinal type data. Bivariate descriptive statistics, specifically zero-order cross-tabulations, allow us to extend our univariate analyses by proposing various hypotheses about a relationship with an independent and dependent variable that is nominal categorical, ordinal categorical or discrete. This non-parametric statistical test works with the basic level of measurement to assess relationships to a certain degree. In crosstabulations, this test is referred to as Chi-Square or denoted as X^2 and is the sample statistic which rejects the null hypothesis.

$$X^2 = \sum \frac{(O_i - E_i)^2}{E_i}$$

X^2 = chi squared
O_i = observed value
E_i = expected value

Crosstabulations

Crosstabulations, sometimes referred to as zero-order crosstabulation is a non-parametric[1] (not dependent on the normal distribution) bivariate statistical test that introduces an independent (X) and dependent (Y) variable and tests a relationship. This test creates a contingency table of the cross-classification of two variables, namely the IV and DV to assess the trends and patterns of the data. This test is highly exploratory and provides a snapshot of how each variable influences the other. This statistical test asks the following questions:

1   Is there a relationship between the Independent and dependent variables?
2   If so, is it statistically significant or not?
3   What is the strength and direction of the relationship?
4   Are there any third unseen variables influencing the original X to Y relationship. Should we test for spuriousness?

These relationships are depicted in a causal model or simple arrow diagram, X denotes the IV and Y is the DV. Additionally, they are written as a formal research question and null (H_0) and research (H_a) hypothesis. For example, a null hypothesis would indicate statistical independency, or no relationship and the alternative or research hypothesis indicates statistical dependency or a relationship between the IV and DV.

Null hypothesis (H_o): There is no statistically significant relationship or statistical dependency between the IV and DV.
Research or Alternate hypothesis (H_a): There is a statistically significant relationship or statistical independency between IV and DV.

Example of an Arrow Diagram or Causal Model in a Bivariate Relationship

Figure 4.1

Requirements for Zero-order crosstabulations

1   Independent, Dependent and Control variable that is nominal or ordinal (categorical or discrete); interval-ratio may be recoded to 'fit' the crosstabulation requirements
2   Have more than 5 cases in each cell (observed count) for a reliable crosstabulation
3   No more than a 4x4 crosstabulation, recode as necessary
4   Zero-order must be statistically significant to elaborate and add a control variable (Z)

In understanding crosstabulation, you must have a thorough understanding of what the IV is and what constitutes the DV. Often students misunderstand the functionality and identification of the IV and DV. To move to the analysis stage, there must be a thorough understanding of this, as well as the levels of measurement that they are. The IV often comes before in time, whereas the DV after. The DV is what is being measured or the outcome or criterion. IVs are often demographic variables, like sex of respondent, age, education, or race, to name a few. The DV are more social phenomena. For example, engagement in domestic violence varies by gender or there is a statistically significant relationship or statistically dependent relationship between gender and domestic violence. Here, generalizability is not the goal but rather the focus is two-fold:

1. To explore how two cross-classified independent and dependent variables significantly relate or not and

1   To provide insights about the patterns and their relationship. Crosstabulations rely on a total of four item statistics to understand the story of the relationship being tested in the contingency table for crosstabulations.

Key Statistics to Report

1 Observed counts or conditional distributions: these are the actual counts or observations for each cross-classified cell; a simple count of observed counts
2 Column percentages by IV are the percentages of each cross-classified cell
3 The non-parametric Chi-Square test (X2) of statistical independence which is the sample statistic that informs of whether the relationship is statistically significant or not; it confirms rejection of the null hypothesis
4 Measures of association provide strength and direction of the relationship

The benefit of crosstabulations is that they are not complicated to analyze, compared to other bivariate statistical tests. The first part of a crosstabulation begins with a valid and missing case summary table. It tells you how many respondents answered the two questions or not. This outlines who answered these survey questions and who did not. Second, the contingency table determines the patterned outcomes. The observed counts or conditional distributions provide the observed counts and column percentages of how many participants from the sample fall in a specific cross-classification cell. Technically, the observed count is the real number of observations in a sample that belong to a specific classification. These counts tell us the highs and lows and do the storytelling of numbers for what is happening in the data in the cross-classification table. It reports the patterns of where they majority or minority lie and reveal interesting data between the IV and DV. These counts also make us aware about variation in the crosstabulation. Did all participants answer in a similar manner? Or do we have variation of responses? What is the difference in cell count across the contingency table? This is important in understanding the outcomes for our sample statistic, Chi-Square (X^2). The crosstabulation's contingency table comes with some core requirements. Prior to running this statistical test, requirements should be carefully reviewed to ensure the correct steps are being taken and the correct statistics being run. See the example below of what a contingency table is like.

	Men	Women	Total
Single	172 (44.2%)	146 (26.9%)	318 (34.2%)
Married	217 (55.8%)	396 (73.1%)	613 (65.8%)
Total	389 (100%)	542 (100%)	931 (100%)

Activity Alert

Try reporting back the statistics from the table and see what the trends and patterns reveal about marital status by gender. Let's do one together: 172 or 44.2% of men are single, you do the next …

The X^2 is the sample statistic used in zero-order crosstabulations. This statistic informs us of whether the IV and DV maintain statistical dependency or not or in other words if the relationship is statistically significant or not. Based on the selected confidence level, 90%, 95% or 99% confidence level and corresponding alphas of 0.10,

0.05, and 0.001. Rejection or fail to reject null hypothesis is based on the calculated X^2 value. Throughout this book, a 95% confidence level is utilized, unless otherwise stated. There are two scenarios possible with the sample statistic:

Scenario A X^2	Scenario B X^2
If the probability of the calculated X^2 or sample statistic is less than 0.05, then the relationship is statistically significant, $p < 0.05$ and the null hypothesis is rejected. A researcher wants to be on this side.	If the probability of the calculated X^2 or sample statistic is less than 0.05, then the relationship is statistically significant, $p < 0.05$ and the null hypothesis is rejected.

The final statistic that is critical in the assessment of the crosstabulation is the measure of association. This measures strength and direction of the nominal or ordinal bivariate relationship. It is not only important to determine statistical significance, but to assess how strong or weak the relationship may be. In statistics there are various measures of association for different levels of measurement. Nominal measures of association are referred to as Type A and range from values between 0 (no association) and 1 (perfect association) and Type B which range from values of 0 to ±1.

Type a: Nominal x nominal	Type b: Ordinal x ordinal
Phi	Gamma
Cramer's V	Yules Q
Contingency Coefficient C	Somers d
Lamba	Tau b

There are many choices of which measure of association to select. However, the most common one for nominal x nominal contingency tables is Cramer's V and for ordinal x ordinal Gamma. These are robust measures and a good selection. A weak relationship may be indicative of an unseen third variable, a control variable, playing a pivotal role in the relationship, while a strong relationship demonstrates that the IV is a strong contender in the relationship. Please note that zero-order crosstabulation must be statistically significant ($p < 0.05$) to elaborate the relationship further in an elaborated crosstabulation. The original x and y relationship must reject the null hypothesis and declare statistical dependency between the IV and DV.

Zero-Order Crosstabulations in SPSS

1   Click on Analyze > Descriptive Statistics > Crosstabulations
2   Move the variables of interest into the Column (IV) and Row Box (DV)
3   Click Cells, select Observed Counts and Column Percentages
4   Click Statistics, select Chi-square and Measure of Association, Cramer's V or Phi for Nominal x Nominal Crosstabulation, Gamma for Ordinal x Ordinal Crosstabulation and Nominal x Ordinal Crosstabulation.
5   Click OK to generate a zero-order crosstabulation.

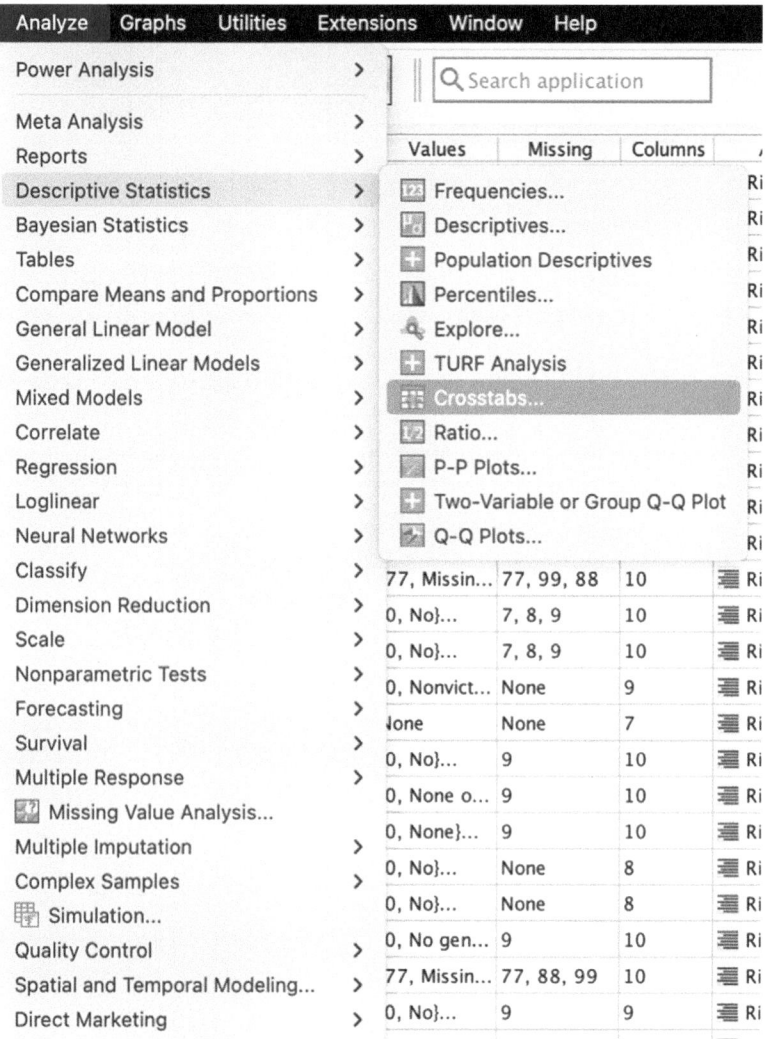

Figure 4.1a Zero-order Crosstabulation in SPSS

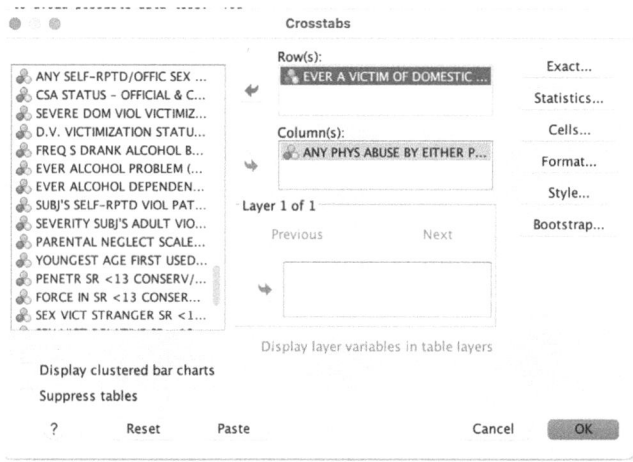

Figure 4.1b Add the IV to the Column and DV to the Row Box

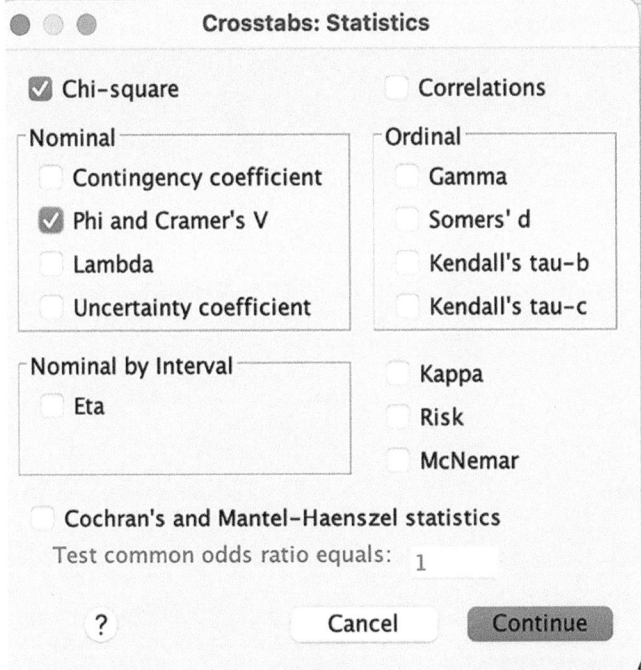

Figure 4.1c Click Statistics and Check-off the Sample Statistic Chi-Square and the Measure of Association in Accordance with Levels of Measurement

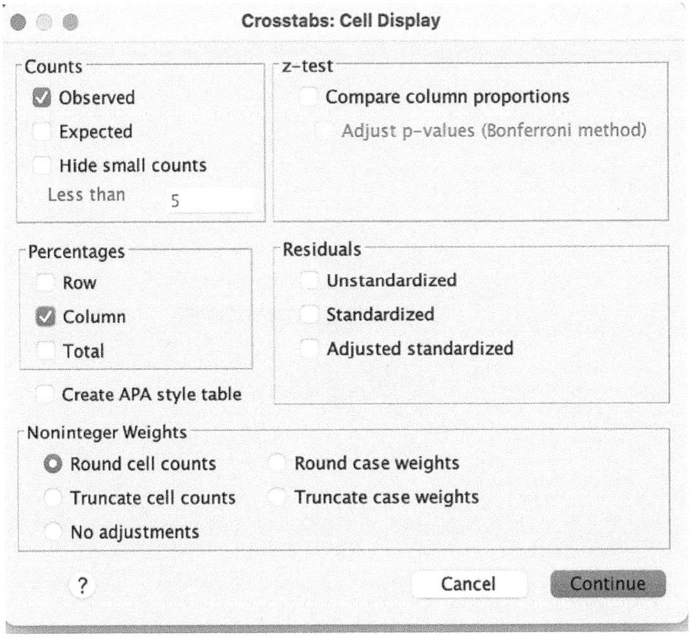

Figure 4.1d Click Cells and Check-off Observed Counts and Column Percentages

Case Processing Summary

	Cases					
	Valid		Missing		Total	
	N	Percent	N	Percent	N	Percent
EVER A VICTIM OF DOMESTIC VIOLENCE * ANY PHYS ABUSE BY EITHER PARENT	173	99.4%	1	0.6%	174	100.0%

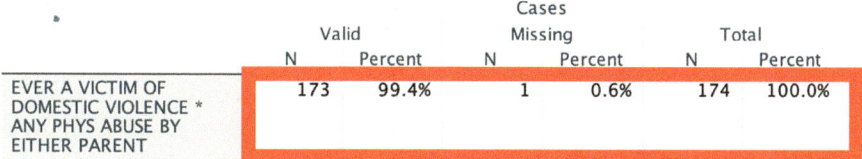

EVER A VICTIM OF DOMESTIC VIOLENCE * ANY PHYS ABUSE BY EITHER PARENT Crosstabulation

			ANY PHYS ABUSE BY EITHER PARENT		
			None	Yes by at least 1 parent	Total
EVER A VICTIM OF DOMESTIC VIOLENCE	No	Count	14	39	53
		% within ANY PHYS ABUSE BY EITHER PARENT	50.0%	26.9%	30.6%
	Yes	Count	14	106	120
		% within ANY PHYS ABUSE BY EITHER PARENT	50.0%	73.1%	69.4%
Total		Count	28	145	173
		% within ANY PHYS ABUSE BY EITHER PARENT	100.0%	100.0%	100.0%

Chi-Square Tests

	Value	df	Asymptotic Significance (2–sided)	Exact Sig. (2–sided)	Exact Sig. (1–sided)
Pearson Chi-Square	5.895[a]	1	.015		
Continuity Correction[b]	4.858	1	.028		
Likelihood Ratio	5.528	1	.019		
Fisher's Exact Test				.024	.016
Linear–by–Linear Association	5.861	1	.015		
N of Valid Cases	173				

a. 0 cells (0.0%) have expected count less than 5. The minimum expected count is 8.58.

b. Computed only for a 2x2 table

Symmetric Measures

		Value	Approximate Significance
Nominal by Nominal	Phi	.185	.015
	Cramer's V	.185	.015
N of Valid Cases		173	

Figure SPSS Output #4.1 Bivariate Zero-Order Crosstabulation of Parental Neglect by Victimization

SPSS Example: Technical and Substantive Interpretation of Zero-Order Crosstabulation

Research Question: *Does any physical abuse by either parent influence if you are a victim of domestic violence?*

Null Hypothesis: There is no statistically significant relationship between the IV, physical abuse by either parent, and the DV, ever a victim of domestic violence.

Research Hypothesis: There is a statistically significant relationship between the IV, physical abuse by either parent, and the DV, ever a victim of domestic violence.

Technical interpretation

Case processing summary: The sample size is 173, with one missing case. There were 99.4% valid cases and 0.6% missing cases (100%).

Contingency table: Approximately 50% (14) who were not abused by at least one parent were not victims of domestic violence, while about 27% (39) who were abused by at least one parent also were not a victim of domestic violence. However, 73.1% (106) who claimed to be abused by either parent identified as being a victim of domestic violence, compared to 50% (14) women that were not abused.

Chi-Square test of Statistical Independence & Measure of Association: The probability of the calculated Chi-Square value of 5.895, 0.015 ($p < 0.05$) is statistically significant and therefore, we reject the null hypothesis. The strength and direction of the relationship is significantly weak (0.185, $p < 0.05$).

Substantive interpretation

Hence, most respondents answered both survey questions, any physical abuse by either parent and ever a victim of domestic violence. Most women who claimed to be abused by either parent identified as a victim of domestic violence. There is statistically dependency amongst the IV and DV. They are indeed related. Further testing should occur with control variables as the original relationship is deemed weak. It is important to note that the representation of individuals for each group was quite different.

Elaborated Crosstabulation (with Control Variable 'Z')

Adding a Control Variable: Elaborated Crosstabulations

The elaborated crosstabulation is an extension of the zero order crosstabulation and tests for spurious relationships for nominal and ordinal data. In other words, it considers how a third variable influences the original statistically significant X to Y or the IV and DV relationship seen in the zero-order crosstabulation. This type of crosstab includes a control variable, Z, into the relationship making it a trivariate or multivariate relationship. Most often in studying associations, one evaluates the relationship between two variables (bivariate), namely the independent and dependent variables. However, a researcher needs to establish if the 'original' relationship is real (or not). Here, the researcher is concerned to know whether there is an inherent link between the independent and dependent variables or whether it is based on an accidental connection with some associated variable. Testing for spurious relationships is an important part of the crosstabulation statistical test. A good researcher must show that the effect is due to the causal variable and not to something else. In short, the researcher must guard against what are called spurious (i.e., association between two variables is shown to be caused by some unseen third variable) relationships through controlling or elaboration of crosstabs. Example: We note from a zero-order crosstab that there is a positive significant association (i.e., statistical dependence) between ice cream sales and deaths due to drowning. The more ice cream sold, the more drowning. Once elaboration is done on this relationship, we find that a third variable, namely, season or temperature, is playing an important role in the relationship. Had we concluded that consumption of ice cream and drowning go hand in hand people would stop going to pools. But in fact, the third variable explains the finding much better. Common Control Variables are: sex, age, race, education, and marital status but can be non-demographic as well.

Once a control variable is added to the relationship various patterns may emerge based on their corresponding Chi-Square values:

1   Replication pattern
2   Spurious pattern
3   Interaction or Conditional pattern

The replication pattern is a 'repeating pattern' of Chi-Square values from the original zero-order crosstabulation to the elaborated crosstabulation. What this means is that the value of the zero-order crosstabulation Chi-Square was statistically significant and continued to be significant for all conditions of the control variable.

The spurious pattern is a 'non-repeating pattern' of Chi-Square values from the original zero-order crosstabulation to the elaborated crosstabulation. What this means is that the value of the zero-order crosstabulation Chi-Square was statistically significant, however with the addition of the control variable all conditions of the control variable became non-significant. The relationship is completely changed with the control variable present, therefore suggesting that the control variable is playing a significant role in the original x and y relationship.

The conditional or interaction pattern is also a non-repeating pattern of Chi-Square values from the original zero-order crosstabulation to the elaborated crosstabulation. However, it is only modified for one condition of the control variable, not both, as seen in the spurious pattern.

All these patterns depict the very important relationship of the third unseen control variable and its relationship to the original zero-order crosstabulation relationship. The role the control variable plays in these relationships speaks volumes to the statistical outcomes and explanations we make about any relationship.

To ensure we understand how elaborated crosstabulations work, let's work through an example that examines the relationship between gender and marital status. The zero-order crosstab Chi-Square revealed a statistically significant relationship. But to

Zero-order crosstab	Emerging pattern	Elaborated crosstab outcome
X and Y are tested $p < 0.05$	Replication Repeating Chi-Square	'Z' or control variable has no impact on the original x and y relationship; All conditions continue to be significant. The original relationships H_o remains rejected; pick alternative controls to test the relationship further
X and Y are tested $p < 0.05$	Spurious Non-repeating Chi-Square	'Z' or control variable has great impact on the original x and y relationship; All conditions change to non-significant. The original relationships H_o remains fails to be rejected; Focus and discuss control(s) variables
X and Y are tested $p < 0.05$	Interaction or conditional partial repeating Chi-Square	'Z' or control variable has partial impact on the original x and y relationship; The original relationships H_o is partially rejected; focus on the condition that changed

ensure that there is no unseen third variable playing a role in the relationship, we elaborate with level of education, high school, and post-secondary education. Now, let's review how these three patterns play out theoretically.

Replication Pattern

The original relationship has been replicated. The Chi-Square tables for each condition of the control are statistically significant and reveal a similar pattern. It is repeating or replicating pattern. Thus, this indicates that the original relationship between gender and marital status did not change with the addition of the control variable of level of education. The original hypothesis that gender influences marital status is supported.

Spurious Pattern

The original relationship has not been replicated. The Chi-Square tables for each condition of the control are not statistically significant and reveal a non-similar pattern. It is not a repeating or not a replicating pattern. Thus, this indicates that the original relationship between gender and marital status has indeed changed. The original hypothesis that gender influences marital status is not supported. Level of education takes precedence.

Interaction or Conditional Pattern

The original relationship has not been replicated. The Chi-Square tables for only *one* condition of the control is not statistically significant and reveal a partially non-similar pattern. It is a partial repeating or replicating pattern. Thus, this indicates that the original relationship between gender and marital status has indeed changed. The original hypothesis that gender influences marital status is only partially supported. Only one condition of the control, the one that is not significant, let's say post-secondary status becomes the focus.

Activity Alert

Can you think of a relationship you would want to elaborate? What would your control variable be and why? Use any data set and build a zero-order crosstab, then add an elaborate the relationship and discuss the findings with the class or peers.

Final Thoughts

The power of control variables is sometimes underestimated but control variables really unravel the mysteries of the relationships being tested. The outcomes of control variables are sometimes very surprising or not. But either way, something is learned about the relationship. Make it a habit to control for variables. Without this step, your analysis is incomplete and therefore your conclusions may be incorrectly presented and inaccurate. Controlling for relevant variables not only provides a more rigorous test of a hypothesis, but it may lead to additional insight if the relationship is found to differ. A theory or social explanation is needed to justify your use of control variables.

A few things to consider when using control variables in crosstabulations:

1   Keep control variable response attributes lean and manageable; most often two response attributes are a good choice. Variables with more than four response

attributes becomes tricky and interpretability, as well as cell count may be compromised. For example, a marital status variable with attributes of single, married, separated, divorced, widowed, can easily recode into married and not married.

2 Ensure your cell counts are greater than 5 people in each cell. If not, the reliability and validity of the crosstabulation is compromised and this must be discussed in your report.

Elaborated Crosstabulations in SPSS

1 Click on Analyze > Descriptive Statistics > Crosstabulations
2 Move the variables of interest into the Column (IV) and Row Box (DV)
3 Move the CONTROL variable to the Layer Box
4 Click Cells, select Observed Counts and Column Percentages
5 Click Statistics, select Chi-square and Measure of Association, Cramer's V or Phi for Nominal x Nominal Crosstabulation, Gamma for Ordinal x Ordinal Crosstabulation and Nominal x Ordinal Crosstabulation.
6 Click OK to generate an elaborated crosstabulation

Key Statistics to Report

1 Case processing summary: Valid and missing cases
2 Contingency table: Observed counts/conditional distributions and column percentages
3 Sample statistic: Pearson Chi-Square X2: Rejection of the Null or not
4 Measure of association: Cramer's V (nominal x nominal), Gamma (ordinal x ordinal, nominal x ordinal)

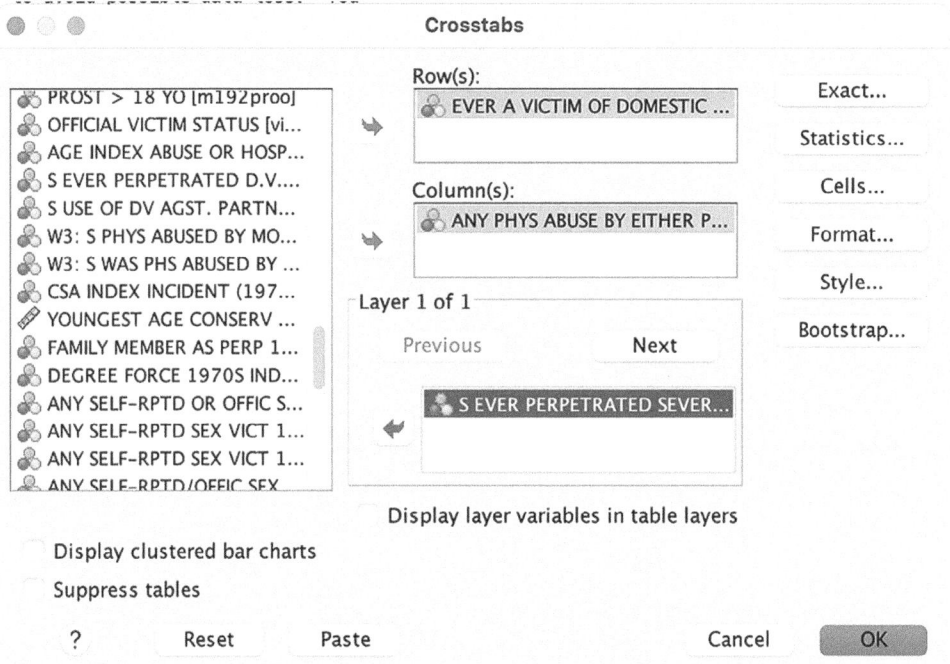

Figure 4.2a Elaborated Crosstabulation in SPSS (Adding a Control, 'Z', to the model) by Adding All Variables, X, Y, and Z to the Designated Boxes with Z in the Layer Box

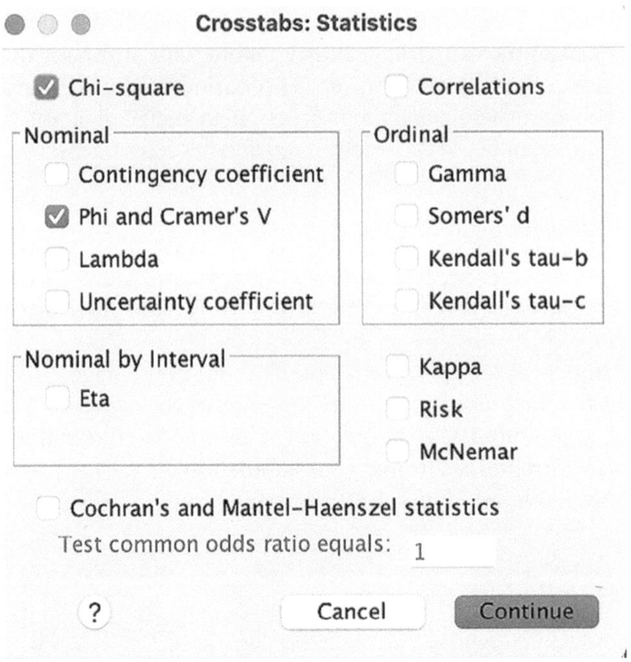

Crosstabs: Statistics

☑ Chi-square   ○ Correlations

Nominal
- ○ Contingency coefficient
- ☑ Phi and Cramer's V
- ○ Lambda
- ○ Uncertainty coefficient

Ordinal
- ○ Gamma
- ○ Somers' d
- ○ Kendall's tau–b
- ○ Kendall's tau–c

Nominal by Interval
- ○ Eta

- ○ Kappa
- ○ Risk
- ○ McNemar

○ Cochran's and Mantel–Haenszel statistics
Test common odds ratio equals: 1

? | Cancel | Continue

Figure 4.2b Click Statistics and Check-off the Sample Statistic Chi-Square and the Measure of Association in Accordance with Levels of Measurement

Crosstabs: Cell Display

Counts
- ☑ Observed
- ○ Expected
- ○ Hide small counts
  Less than 5

z-test
- ○ Compare column proportions
- ○ Adjust p-values (Bonferroni method)

Percentages
- ○ Row
- ☑ Column
- ○ Total

○ Create APA style table

Residuals
- ○ Unstandardized
- ○ Standardized
- ○ Adjusted standardized

Noninteger Weights
- ● Round cell counts
- ○ Truncate cell counts
- ○ No adjustments
- ○ Round case weights
- ○ Truncate case weights

? | Cancel | Continue

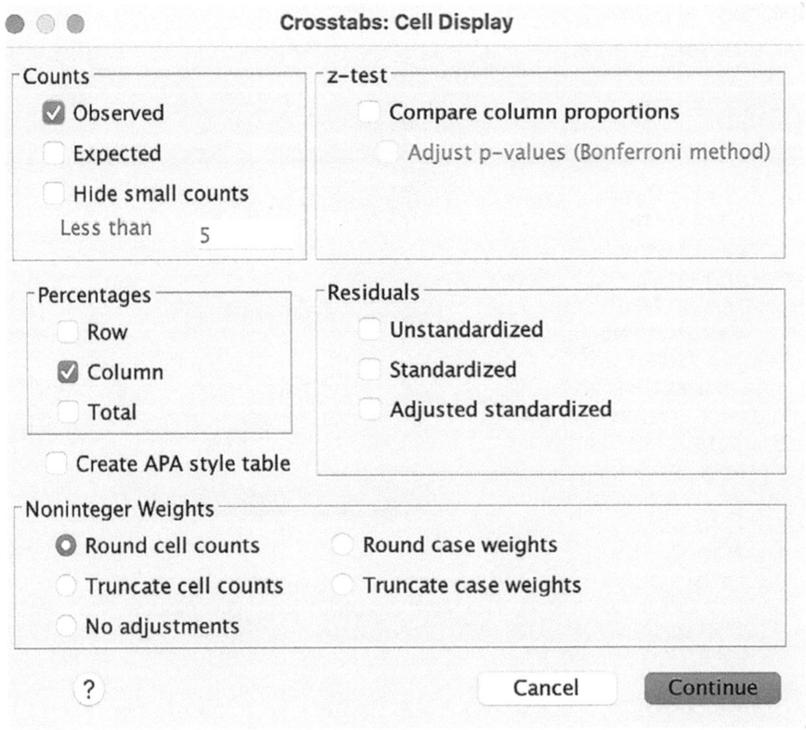

Figure 4.2c Click Statistics and Check-off the Sample Statistic Chi-Square and the Measure of Association in Accordance with Levels of Measurement

Case Processing Summary

	Cases					
	Valid		Missing		Total	
	N	Percent	N	Percent	N	Percent
EVER A VICTIM OF DOMESTIC VIOLENCE * ANY PHYS ABUSE BY EITHER PARENT * S EVER PERPETRATED D.V.	173	99.4%	1	0.6%	174	100.0%

EVER A VICTIM OF DOMESTIC VIOLENCE * ANY PHYS ABUSE BY EITHER PARENT * S EVER PERPETRATED D.V. Crosstabulation

					ANY PHYS ABUSE BY EITHER PARENT		
S EVER PERPETRATED D.V.					None	Yes by at least 1 parent	Total
No	EVER A VICTIM OF DOMESTIC VIOLENCE	No		Count	10	25	35
				% within ANY PHYS ABUSE BY EITHER PARENT	83.3%	56.8%	62.5%
		Yes		Count	2	19	21
				% within ANY PHYS ABUSE BY EITHER PARENT	16.7%	43.2%	37.5%
	Total			Count	12	44	56
				% within ANY PHYS ABUSE BY EITHER PARENT	100.0%	100.0%	100.0%
Yes	EVER A VICTIM OF DOMESTIC VIOLENCE	No		Count	4	14	18
				% within ANY PHYS ABUSE BY EITHER PARENT	25.0%	13.9%	15.4%
		Yes		Count	12	87	99
				% within ANY PHYS ABUSE BY EITHER PARENT	75.0%	86.1%	84.6%
	Total			Count	16	101	117
				% within ANY PHYS ABUSE BY EITHER PARENT	100.0%	100.0%	100.0%
Total	EVER A VICTIM OF DOMESTIC VIOLENCE	No		Count	14	39	53
				% within ANY PHYS ABUSE BY EITHER PARENT	50.0%	26.9%	30.6%
		Yes		Count	14	106	120
				% within ANY PHYS ABUSE BY EITHER PARENT	50.0%	73.1%	69.4%
	Total			Count	28	145	173
				% within ANY PHYS ABUSE BY EITHER PARENT	100.0%	100.0%	100.0%

Figure SPSS Output #4.2 Elaborated Crosstabulation Testing for Spurious Relationships With The Control Variable, S Ever Perpetrated Domestic Violence

Chi-Square Tests

S EVER PERPETRATED D.V		Value	df	Asymptotic Significance (2-sided)	Exact Sig. (2-sided)	Exact Sig. (1-sided)
No	Pearson Chi-Square	2.828[c]	1	.093		
	Continuity Correction[b]	1.810	1	.178		
	Likelihood Ratio	3.105	1	.078		
	Fisher's Exact Test				.177	.086
	Linear–by–Linear Association	2.778	1	.096		
	N of Valid Cases	56				
Yes	Pearson Chi-Square	1.316[d]	1	.251		
	Continuity Correction[b]	.600	1	.439		
	Likelihood Ratio	1.174	1	.279		
	Fisher's Exact Test				.268	.212
	Linear–by–Linear Association	1.305	1	.253		
	N of Valid Cases	117				
Total	Pearson Chi-Square	5.895[a]	1	.015		
	Continuity Correction[b]	4.858	1	.028		
	Likelihood Ratio	5.528	1	.019		
	Fisher's Exact Test				.024	.016
	Linear–by–Linear Association	5.861	1	.015		
	N of Valid Cases	173				

a. 0 cells (0.0%) have expected count less than 5. The minimum expected count is 8.58.

b. Computed only for a 2x2 table

c. 1 cells (25.0%) have expected count less than 5. The minimum expected count is 4.50.

d. 1 cells (25.0%) have expected count less than 5. The minimum expected count is 2.46.

Symmetric Measures

S EVER PERPETRATED D.V.			Value	Approximate Significance
No	Nominal by Nominal	Phi	.225	.093
		Cramer's V	.225	.093
	N of Valid Cases		56	
Yes	Nominal by Nominal	Phi	.106	.251
		Cramer's V	.106	.251
	N of Valid Cases		117	
Total	Nominal by Nominal	Phi	.185	.015
		Cramer's V	.185	.015
	N of Valid Cases		173	

SPSS Example: Technical and Substantive Interpretation

Elaborated Crosstabulations or First Order Crosstabulation

Research Question: *Does any physical abuse by either parent influence if you are a victim of domestic violence, while controlling for women perpetrated domestic violence?*

Null Hypothesis: There is no statistically significant relationship between the IV, physical abuse by either parent, and the DV, ever a victim of domestic violence, while controlling for women perpetrated domestic violence?

Research Hypothesis: There is a statistically significant relationship between the IV, physical abuse by either parent, and the DV, ever a victim of domestic violence, while controlling for women perpetrated domestic violence?

Technical interpretation

Case processing summary: The valid cases for the elaborated crosstabulation are 173. The response rate for the cross-classified variables were about 99% with about 1 missing case.

Contingency Table: Roughly, 83% (10) of non-victimized women, compared to about 17% (2) that claimed to be victims of domestic violence, were not physically abused by either parent, did not ever perpetuate the abuse. In contrast, about 57% (25) non-victimized women who were physically abused by either parent, did not ever perpetuate the abuse, while 43% (19) who were physically abused by either parent, did perpetuate domestic violence. Alternatively, of those women that perpetrated domestic violence, 25% (4) claimed no physical abuse, as well as were not ever victims of domestic violence; 75% (12) of those women that perpetuated the abuse had no counts of abuse by parents but were victims of domesticated abuse. In contrast, of those women that claimed to perpetuate domestic violence, approximately, 14% (14) indicated that they were abused by either one parent but did not identify as victims of domestic violence, compared to 86% (87) that were abused by one parent, claimed to be a victim of domestic violence.

Chi-Square test of statistical independence and Measure of association: The probability of the calculated Chi-Square values for each condition of the control, those that did perpetuate domestic violence versus those that did not, have values of 2.828, 0.093 ($p > 0.05$) and 1.316, 0.251, are both not statistically significant. We fail to reject the null hypotheses. Indicative of a 'spurious' relationship. The original relationship is altered and becomes non-significant with the addition of the control variable and the relationship remains weak. The focus of the analysis becomes the control variable.

Substantive Interpretation

In summary, the results of the elaborated crosstabulation have shown that many women who had at least one parent abuse them were victims of domestic violence. When a third unseen variable is added to the model, namely, perpetuate domestic violence, the original relationship is modified completely. Perpetuating domestic violence has shown to influence the original relationship of abused by either parent and ever a victim of domestic violence. Indeed, your past relationships with your folks coupled with being a domestic violence victim or survivor, may perpetuate domestic violence or not. Several social explanations can heed to this finding. The elaborated crosstabulation must be taken by caution as there are some cells that have counts of less than five, for this reason reliability of the crosstabulation is questionable. We can take the results for what it's worth. Also, certain groups have larger representation, than others. This may imbalance the statistical outcomes.

Final Thoughts

The descriptive statistic of zero-order and elaborated crosstabulations incorporates testing relationships using nominal and ordinal data. This is a basic bivariate or multivariate statistical technique that expands univariate analyses in so many ways. Using a non-parametric test, like Chi-square, one can assess whether or not a DV varies with an IV and how strong that relationship is or not. This statistical test allows for one to come to a better understanding of cross-classifying two variables together and explore the outcomes, as well as make informed decisions about which control variables may be playing a role in that relationship. While generalizability is not the goal, nor are inferences from a sample to a population, crosstabulations allows us to see how participants responded and how the data varies. Crosstabulations may help build relationships for inference based statistical testing, especially for those outcomes that deem statistical significance.

Keywords and Definitions

Parametric vs. nonparametric test	Usually these are associated with population parameters. These tests are most often inference-based, and their main objective is to infer findings from a sample to a population parameter, based on the normally distributed population distribution, interval-ratio continuous data and large sample size. Non-parametric tests are not bound to the normal distribution because they work with nominal and ordinal type data.
Bivariate test	A test that examines two variables, most often an IV and DV.
Crosstabulations	sometimes referred to as the zero-order crosstabulation is a non-parametric[2] (not dependent on the normal distribution) bivariate statistical test that introduces an independent (X) and dependent (Y) variable and tests a relationship.
Observed counts or conditional distributions	These are the actual counts or observations for each cross-classified cell; a simple count of observed counts.
Column percentages	The percentages of each cross-classified cell.
Chi-Square test (X2) of statistical independence	A non-parametric test that is the sample statistic that informs of whether the relationship is statistically significant or not; it confirms rejection of the null hypothesis
Measures of association	These provide strength and direction of the relationship and ranges from 0 to ±1. There are different measures of association for nominal x nominal variables and ordinal x ordinal variables. Examples include Cramer's V and Gamma respectively.
Replication pattern	A 'repeating pattern' of Chi-Square values from the original zero-order crosstabulation to the elaborated crosstabulation. What this means is that the value of the zero-order crosstabulation Chi-Square was statistically significant and continued to be significant for all conditions of the control variable.

| Spurious pattern | A 'non-repeating pattern' of Chi-Square values from the original zero-order crosstabulation to the elaborated crosstabulation. What this means is that the value of the zero-order crosstabulation Chi-Square was statistically significant, however with the addition of the control variable all conditions of the control variable became non-significant. The relationship is completely changed with the control variable present, therefore suggesting that the control variable is playing a significant role in the original x and y relationship. |
| Conditional or interaction pattern | Also a 'non-repeating' pattern of Chi-Square values from the original zero-order crosstabulation to the elaborated crosstabulation. However, it is only modified for one condition of the control variable, not both, as seen in the spurious pattern. |

Test Your Knowledge

1   A _____ relationship indicates that when a control variable is added and only one condition of the control impacts the original relationship.

   a   Interaction or Conditional
   b   Spurious
   c   Replication
   d   Chi-Square
   e   None of the above

2   The crosstabulation does the following:

   a   Tests a bivariate relationship with an IV and DV using a contingency table
   b   Engages in single variable analyses
   c   Uses Chi-Square as the sample statistic
   d   Tests the level of measurement of all variables
   e   Replicates the science of all variables

3   The _____ is the sample statistic of the crosstabulation and elaborated crosstabulation and tells the researcher about statistical independency or dependency of the IV and DV.

   a   t-test
   b   Median
   c   F-test
   d   Chi-Square
   e   Measure of Associations

4 Crosstabulations test a relationship. The relationship is statistically significant, and the null hypothesis is rejected when:

   a $p < 0.05$
   b $p > 0.05$
   c   Cramer's V is significant
   d   When the IV is powerful
   e   When a control variable is added to the model

5 Chi-Square is a referred to as a _____ test.

   a   Population
   b   Sample
   c   Parameter
   d   Parametric
   e   Non-parametric

6 The university has hired you to investigate a bivariate relationship. They are measuring diversity, inclusion, and equity (DIE) and exploring factors like race, gender, and faculty rank. What is the IV and DV identified in this research?

   a   IV: DIE and DV: race, gender, faculty rank
   b   IV: race, gender, faculty rank and DV: DIE
   c   These are all control variables
   d   There are no IVs or DVs
   e   a and b

7 _____ provide strength and direction of the relationship and ranges from 0 to ±1. There are different measures of association for nominal x nominal variables and ordinal x ordinal variables. Examples include Cramer's V and Gamma respectively.

   a   Chi-Square
   b   Measures of association
   c   Observed counts
   d   Replication pattern
   e   Conditional distributions

8 Observed counts or Conditional distributions are the actual counts or observations for each cross-classified cell; a simple count of observed counts.

   a   True
   b   False

9   Tiffany ran an elaborated cross-tab. The results of the test are below:

Case Processing Summary

	Cases					
	Valid		Missing		Total	
	N	Percent	N	Percent	N	Percent
Smoked every day for 7 days in a row * Gender of Youth 1=Female	2357	12.4%	16661	87.6%	19018	100.0%

Smoked every day for 7 days in a row * Gender of Youth 1=Female Crosstabulation

			Gender of Youth 1=Female		Total
			Male	Female	
Smoked every day for 7 days in a row	Yes	Count	438	553	991
		% within Gender of Youth 1=Female	38.3%	45.6%	42.0%
	No	Count	706	660	1366
		% within Gender of Youth 1=Female	61.7%	54.4%	58.0%
Total		Count	1144	1213	2357
		% within Gender of Youth 1=Female	100.0%	100.0%	100.0%

Chi–Square Tests

	Value	df	Asymptotic Significance (2–sided)	Exact Sig. (2–sided)	Exact Sig. (1–sided)
Pearson Chi–Square	12.885[a]	1	<.001		
Continuity Correction[b]	12.587	1	<.001		
Likelihood Ratio	12.904	1	<.001		
Fisher's Exact Test				<.001	<.001
Linear–by–Linear Association	12.880	1	<.001		
N of Valid Cases	2357				

a. 0 cells (0.0%) have expected count less than 5. The minimum expected count is 480.99.
b. Computed only for a 2x2 table

Symmetric Measures

		Value	Approximate Significance
Nominal by Nominal	Phi	-.074	<.001
	Cramer's V	.074	<.001
N of Valid Cases		2357	

Figure 4.2

What information does the above zero-order crosstabulation convey?

a   The relationship is testing whether or not smoking daily varies by minority status
b   An increase in percentage of men and women did not smoke daily, compared to those that did
c   The probability of the calculated Chi-Square test statistic was *not* statistically significant at the 95% confidence level.
d   An elaboration cannot take place due to the Chi-Square outcome
e   None of the above

10 Elaborated crosstabs explore trends and patterns of a control variable and compare it to the original relationship by examining the pattern of what statistic?

a   Gamma
b   Cramer's V
c   Chi-Square
d   Frequencies
e   Factor analysis

Notes

1 Non-parametric tests do not follow the normal distribution and are more likely to be used with ordinal or nominal data.
2 Non-parametric tests do not follow the normal distribution and are more likely to be used with ordinal or nominal data.

Part 2

Inferential Statistical Analyses Using Hypothesis Testing, Comparison of Means Tests and Average Differences

This second portion of the textbook explores inferential statistics, with a specific focus on the comparison of means or average differences using various sophisticated and slightly advanced statistical analyses. After reading and reviewing this chapter you will be able to develop an understanding of inferential statistics and comparison of means tests, like t-test, One-Way ANOVA, Two Way Factorial ANOVA, ANCOVA, and MANOVA. Examinations of average or mean differences are essential in research and data analysis as it allows us to better understand group differences and make meaningful conclusions about data. This chapter teaches you about the different types of statistical analyses that test for mean differences and, more importantly, how to choose which test and analyze it both technically and substantively.

5 Introduction to Inferential Statistics and Understanding the Nuances of Inference based Testing and Generalizability of Data

Unlike descriptive statistics, comparison of means tests go one step further and begin to explore the nature of population and sampling distributions. These tests are notably different and fall under the inferential test umbrella of hypothesis testing and comparison of means tests. Here, relationships that are found in the social world are navigated and investigated in a variety of ways using unique statistical analyses, each one with its own merits. These data analytical tests allow you to compare average differences or mean differences between two groups or more than two groups, depending on the test that is selected, against a DV(s). Every test here comes with its own prerequisites, conditions, and assumptions. Assumptions in inferential statistical testing is a very critical component of data analysis and should not be taken lightly. Ensuring that all assumptions are met is key to running these statistical tests. When assumptions are the met and not violated the outcome of data becomes reliable and valid; there is enhance validity in the findings and conclusions made; Type 1 and 2 errors are minimized; statistical power of the test increases; less bias in estimates; and confidence in reporting results increases. Inferential statistical tests are highly informative and powerful and go beyond description of variables. Hypothesis testing is a statistical analysis used to make inferences about a population parameter from a sample.

Key Characteristics of Hypothesis Testing for Comparison of Means Tests

Null hypothesis	A statement of no average differences or mean differences
Alternative hypothesis	A statement of average differences or mean differences
Sample test statistic	The sample statistic calculated from the sample data to assess the null hypothesis; F-statistic or F-ratio
Significance level and p-value	Tells the probability of rejecting the null hypothesis or calculated sample or test statistic, commonly set at an alpha value of 0.05 (95% confidence level); $p < 0.05$, reject the H_o and have significant average differences
Decision rule	The calculated test or sample statistic results in a decision to reject the H_o or fail to reject the H_o.

While the scope of this book can't cover everything, it truly tries to discuss the core and most popular comparison of means tests that are used by social science majors and teaches how to write about them once the output is done. All the tests discussed here compare means or averages, how they do this differs with each test. The key statistic to report in the interpretations are the means or averages. However, each test brings its own unique interpretation to the table and must be understood slightly differently. While the likenesses amongst the tests are considerable, their unique statistical testing abilities are what differentiates each test. At the end of this part, you should have good grasp of what hypothesis testing statistical analyses techniques are like and be able to make an informed decision of what test fits the data you set out to analyze.

DOI: 10.4324/9781003215691-7

Independent samples t-test	One-way ANOVA	Two-way ANOVA	Multivariate ANOVAs
Compares means for a dichotomous IV and interval-ratio DV	Compares means for a non-dichotomous IV and interval-ratio DV	Compares means for two non-dichotomous IVs and interval-ratio DV	Compares means for non-dichotomous IVs coupled with a Covariate and interval-ratio DVs
Two groups	More than two groups	More than two groups	More than two groups
Bivariate	Bivariate	Multivariate	Multivariate
Average differences	Average differences	Average differences	Average differences
Assumptions	Assumptions	Assumptions	Assumptions
Example: Gender differences and test scores	Racial differences and test scores	Racial differences and degree program and test scores	Racial differences and degree program, with covariate: employment and test scores and hours of sleep
No post-hoc testing	Post-hoc testing	Post-hoc testing	Post-hoc testing
Cohen's d	Eta^2	Partial Eta^2 and adjusted R^2	Partial Eta^2 and adjusted R^2

In inferential statistics one of the key issues for students and others is to understand when to use each statistical analysis or technique. There is no hard or fast rule. But there are three main points that inform your decision:

1   The research question or objectives and research hypotheses
2   The nature of your data and the level of measurement of variables and type of variables
3   The # of IVs and DVs to be analyzed

These are three essential questions to ask yourself at the time of data analysis. If you have the answers to these questions, then the decision regarding statistical testing can be made easily. As you navigate these chapters, pay attention to the data requirements of each test. Each level of measurement of each variable determines your statistical pathway. Understanding levels of measurement and types of variables lies at the heart of statistical testing and is so very important. Thus, if you are forgetting, please go back to the earlier chapters and review these levels of measurement.

Activity Alert

Discuss the similarities and differences of each type of 'mean difference' test. Use the table to help you out.

The quality of data is an important aspect of inferential statistics-based testing. The pre-analysis data screening is vital to ensure that quality of data is not compromised and is relatively critical in conducting.

The first important aspect that speaks to quality of data collected is accuracy of data collected. Inaccurate data results in flawed conclusions. Often, with large samples, descriptive statistics is run to checks for data accuracy.

The second is the handling of missing cases in a data set. Random or non-random missing data impacts the quality of data and therefore needs to be handled properly prior to data analysis.

Third, any extreme score or outlier deems problematic as well and may alter results in a negative manner, thus impacting the quality of statistical testing. Outliers or extreme scores, in either direction, needs to be methodically taken care of or removed from data sets accordingly, such that the statistical tests are not affected, and the results or findings are high quality outcomes.

Lastly, as statistical tests become more complex in nature, there are numerous assumptions (i.e., linearity, homoscedasticity, normality, to name a few), especially at the inference-based level that must be tolerated or met for quality results. The upcoming chapters discuss various statistical tests that are 'comparison of means' tests, like, Independent samples t-test, One-way ANOVA, Two-way factorial ANOVA, ANCOVA, MANOVA and a brief discussion of MANCOVA.

This section of the book focuses on the second and more sophisticated branch of statistics, namely, Inferential statistics. Sometimes, the social world demands a different and unique perspective to navigate the questions we may have, and we get to those answers using inferential statistics-a different mindset and approach. These statistics again are interdisciplinary and not just bound to the social sciences. Their use is seen across disciplines; the difference is the variables that are being tested. Education, business, healthcare, science, political science, psychology, sociology, criminology, technology, to name some, make use of inference-based statistics. These advanced statistical techniques allow us to know about 'average differences' or 'predictability' of two variables, namely IV and DV, sometimes control variables or more. For example, we may want to learn about 'average' or 'mean' differences between GPA and gender, or you may want to go one step further and do some 'predictive' analysis. Regardless of discipline, there are always hypotheses that are willing to be tested in some best way. We can go beyond univariate and descriptive statistics levels to understand nuanced relationships. These are more advanced methods of data analysis that are heavily immersed in the concept of hypothesis testing of mostly two variables and IV and DV, but sometimes can follow multivariate testing.

The major difference between the two branches of statistics, as discussed earlier, is that Inferential statistics are statistical testing procedures that allow for the generalizability of data. They infer from a sample to a population. Through hypothesis testing procedures, these statistics allow to test relationships with the highest level of measurement of variables-interval-ratio continuous or its equivalent (i.e., scaling, or dummy coding).

Descriptive statistics	Inferential statistics
1 Describes trends and patterns of the data	1 Hypothesis testing and making predictions
2 Data reduction technique	2 Comparison of means, associations and predictability
3 Highly exploratory in nature	
4 No generalizability of findings	3 Inference-based technique
	4 Tests Generalizability of findings by ensuring sample statistic equates population parameters

Here, we work with samples and infer to populations because we cannot research everyone. Quantitative researchers work with subset of the population which is the sample that should be representative of the population. Through probabilistic random sampling procedures or equal probability selection methods (EPSeM), representative samples are achieved and generalizability of data or inferences from a sample to a population can be made. We do not have means or feasibility or money to study everyone, thus working with a subset of a population, the sample, makes it highly practical and possible. Smaller samples speak to the population parameters which are characteristics of the population. For example, you may want to study all university students, or all homeowners, or all youth criminals to understand the relationships but need to work with a subset of each of these populations.

Randomized probability samples	*Non-randomized non-probability samples*
Simple random sampling Systematic sampling Stratified sampling Multi-stage cluster sampling	Convenience sampling

In statistics, there are different methods to get a randomized sample. The four common types of random sampling methods that allow us to achieve randomized samples that represent the population parameters are:

1   Simple random sampling
2   Systematic sampling
3   Stratified sampling and
4   Multi-stage cluster sampling

First, Simple random sampling. This sampling method is by far the most common and ensures that everyone in the population has an equal chance of being selected.

Second, Systematic sampling is also a form of simple random sampling in which individuals or units are selected from sampling frame or list and a sampling interval is calculated by using the following formula:

$$Sampling\,interval \; = \; \frac{population}{sample}$$

For instance, if the population is 10,000 and the sample size to achieve is 1,000. The sampling interval is calculated to be every 10th individual or unit.

Third, Stratified sampling randomizes population units and individuals by certain strata or categories, like gender, race, rural, urban, and then randomly selects from those strata or categories.

Lastly, Multi-stage cluster sampling, used when a sampling frame or list is not available or too costly or untimely to produce. When population lists are hard to achieve this sampling is used. This type of sampling utilizes many stages or clusters of sampling, and often covers wide geographic areas in which units or individuals are drawn from. A particular city would be the large cluster, the second cluster might be a specific suburb, within that suburb there would be smaller clusters of neighborhoods and within that neighborhood houses would be randomly selected.

Sometimes randomized samples are not achievable for various reasons. Perhaps there is not enough time or there are cost issues or there is no sampling frame or list available. Thus, in those instances researchers target non-probabilistic sampling methods which are non-randomized methods for sampling. Most popular are convenience sampling methods. In this method of sampling, participants are selected based on easy recruitment and accessibility. There are no huge costs associated with this. The sample is selected because of comfort, cost, accessibility, and ease. A prime example of this is recruiting participants from university classes. These participants are easy targets for doing research. Sometimes, positive reinforcement can be introduced. For example, one bonus mark towards the final grade will be given to those that participate. This type of incentive draws in crowds of students and a decent sample is achieved in a timely manner. However, this type of sampling does not come without limitations. There is a major disadvantage to having this type of sample and that is the lack of generalizability of data and findings and perhaps a smaller sample size. Other limitations include sampling bias and sampling error. Sampling bias and error occurs when participants are not pooled from a randomized sample and there is no equal representation of selection; as well sample size is compromised. Thus, estimations of population parameters may be done in error. Extreme caution and careful consideration need to be taken when dealing with these types of sampling. Response rates of surveys are greater with randomized samples. Remember, that a high response rate is genuinely preferred as it enhances representativeness of the sample.

Activity Alert

Try to create real life examples of how you would sample any population of your choice. Use the four random sampling methods to establish your case and explain how the sample is achieved.

Remember, for samples to equate population parameters, the sample must reflect the exact proportions of the population. Consequently, if a university population has 60% men, 30% women and 10% LGBTQIAS+, the sample that is representative of this should match these population parameters and have similar sample characteristics. This is critical to randomized sampling. In this way, inferences and generalizability can be made accurately.

The heart of inferential statistics stems from three key distributions and the Central Limit Theorem:

Population distribution	The sample distribution	The Sampling distribution
True parameters of the population	Sample characteristics	A theoretical or non-empirical distribution of all possible sample outcomes of a given statistic for a certain sample size (N). The sampling distribution contains the sample statistics calculated from sample(s). This distribution allows us to make inferences about population parameters

The Central limit is an essential concept of inferential statistics and allows statisticians to make inferences about. This theorem states, that if repeated samples of size N are drawn from any population, with mean ц and standard deviation б, as N becomes larger, the sampling distribution of sample means will approach normality, with mean ц and standard deviation of б/√N. Whenever sample size is large, it can be assumed that the sampling distribution is normally distributed. A large representative sample size results in normal-like distributions, even if the population from which samples are drawn from are not normally distributed. Normality is based on the properties of the normal curve. The normal curve hugely dictates inference statistics. It is one of the key assumptions that must be met in most hypothesis testing methods and confidence levels, confidence interval estimations and eventually statistical significance. Basically, the Central limit theorem justifies the use of parametric statistical testing procedures.

What are confidence levels? All quantitative researchers must have established confidence levels when 'doing statistics'. Confidence levels must be established to determine statistical significance of outcomes. Considering research design, sampling, errors, type of data produced the confidence levels are set to determine the researcher's 'willingness to be wrong'. Confidence levels range from 90%, 95% and 99%, with corresponding alphas of 0.10, 0.05, and 0.001. These simply mean that the researcher is willing to be wrong by a certain percentage. For example, a 95% confidence interval suggest that the researcher is willing to be wrong only 5% of the time or that 95% of the constructed intervals would contain the population value according to the method's success and 5% would not. Statistical significance is not being 100% confident and confirms the level of risk you are willing to make and take in rejecting the null hypothesis. Basically, does the finding or result in a sample likely represent a true effect or is it random chance? In short, confidence levels provide a simple way to measure how uncertain we are about the sample statistic being calculated. The choice of confidence levels truly depends on the research design, certainty, and precision of statistical inference. Confidence levels should most always be reported in any writing as it tells us the amount to risk a researcher is willing to take with their results. The greater the confidence level, the greater the certainty of results and reliability of data outcomes.

Confidence level	Alpha level	Risk	p-value
90%	10% or 0.10	10% risk of being wrong	0.10
95%	5% or 0.05	5% risk of being wrong	0.05
99%	1% or 0.01	1% risk of being wrong	0.01

This brings us to the next critical discussion of statistical significance. With confidence levels, comes the concept of statistical significance which is especially important in inference-based testing. The two are inter-related. Statistical significance basically states that any effect or difference in results is real and not based by random chance alone and focuses heavily on the p-value. The p-value is the key marker of statistical significance. A small p-value is indicative of statistical significance. Statistical significance is closely tied to hypothesis testing and is a predetermined significance level by the researcher(s). It basically is the core framework of statistical hypothesis testing. Statistical outcomes are best when their sample statistic is statistically significant. This means that there is a relationship or there are average differences or predictions or associations in place amongst the IV and DV. However, if results are not statistically significant this tells us something just as important. A non-significant outcome also reveals important information about the data. Always remember that. Do not be quick to reject any non-significant outcome. That also warrants some attention as it also tells a story of its own and it is important to understand the non-significant nuances of relationships and effects. Non-significant outcomes also provide great information about the social world and what may not work, thus just as important to discuss and explain.

No research is perfect, or error free. There will always errors of sorts in the way the research has been done. In inferential statistics we should avoid Type 1 and 2 errors. These errors result in faulty conclusions or remarks about an outcome of a statistical test. The probability of increasing these errors is correlated with sample size. Type 1 (alpha error): Rejection of the null hypothesis when the null is true; *False-Positive* – you decide treatment works when it doesn't really; and Type 2 (beta error): Failure to reject the null hypothesis when the null is false; you decide treatment doesn't work when really it does.

Statistically significant	Not statistically significant
The probability of the calculated test statistic is less than 0.05, $p < 0.05$; the null hypothesis is rejected.	The probability of the calculated test statistic is greater than 0.05, $p > 0.05$; fail to reject the null hypothesis

In quantitative research, there will always be some error present. No research will yield 100% perfect outcomes and no statistical test will give you 100% accuracy. We need to be cognizant of the strengths and weaknesses of the research design and the statistical outcomes. All researchers need to consider the implications of each type of error and be able to strike a good balance between statistical rigor and practical significance and the statistical test being utilized.

While there are many statistical tests that belong in the inferential tests category. In this part of the textbook, only popular hypothesis testing statistical techniques that compare means, like independent samples t-test, One-way ANOVA, Two-way factorial ANOVA, and ANCOVA and MANOVA with MANCOVA are discussed. All these tests have one common theme: comparison of means through hypothesis testing; how they do it varies and thus the interpretations and attention to detail varies. These tests work with nominal and ordinal IVs with varying response attributes and an interval-ratio continuous DV.

To ensure quality of data has been met, dealing with missing data, outlier and various assumptions must be met always. Statistical pre-screening data is an important process prior to engaging in any inference-based test. This ensures your results are valid and reliable.

Type 1 error (Alpha)	Type 2 error (Beta)
Occurs when you reject a true null hypothesis	Occurs when you fail to reject a false null hypothesis
Conclude that there is a significant effect or difference when in truth, there is no effect or difference in the population; rejecting the null hypothesis, where no effect is present.	Conclude that there is no significant effect or difference when in truth, there is an effect or difference in the population; failing to reject the null hypothesis, when there is an effect present
False positive or alpha error	False negative or beta error
Correct this by inflating the alpha value from 0.05 to 0.01; use statistical corrections; replicate study	Correct this by increasing sample size; choose a powerful statistical test; control for variables

In the end, there are several advantages of using inferential statistics to tell the story of data. Below are some key advantages.

Pre-screening data	Explanation
Data cleaning and coding	Address missing values or any arbitrary values and ensure variables are correctly coded. This ensures accuracy of data
Descriptive statistics	Run Descriptive Statistics to assess the trends and patterns of the data through counts, typical average or score and spread of the distribution
Data distribution and shape of distribution	Examine distributions for skewness or kurtosis, especially check for normality of data
Extreme scores or outlier detection	Be quick to identify any outliers that my skewing data in a negative manner
Variable relationships	Explore the association of variables and how they co-vary. Assess if each variable brings it unique variance to the model(s) being tested and no instances of multicollinearity are present
Data transformation	To normalize data, engage in any data transformations; engage in any data modifications that design variables for a particular inference-based test
Sample	Always be critical of the sample and evaluate its representativeness of the population. Biases may impact inferences made

Advantages of Using Inferential Statistics

Generalizability	Allows for generalization from a subset of the population, the sample to a larger population
Efficiency	Statistical methods for larger populations that are far more efficient than descriptive statistics
Hypothesis Testing	Facilitate hypothesis testing of comparison of means or average differences using t-tests and ANOVA family statistical methods
Prediction	Facilitate predictions of large populations using regression models
Comparisons	Enables comparisons of groups by assessing average or mean significant differences of the IV(s) and DV(s)
Patterning	Inferential Statistics analyze patterns from samples to infer to a larger population
Decision Making	Assist in making effective decisions in treatments and groups
Cost and Time Effective	Allows to work with a sample and make inferences accordingly
Policy Implications	Statistical outcomes play a vital role in shaping policies and interventions

Final Thoughts

The intent of this chapter was to familiarize you with the main tenets of inferential statistics and the core principles that make these statistics worthwhile. It is important to grasp the core ideas to move on to the next chapters.

Here are some key steps that can be followed as you navigate these chapters:

Key Steps in Comparison of Means Tests

Formulate hypotheses	*A statement of no average differences or mean differences* *A statement of average differences or mean differences*
Select appropriate test or sample statistic	Select appropriate sample statistic based on the research design, nature of data, and research question or objectives
Collect and analyze data	Collect survey data and analyze data using SPSS or any other software
Determine significance	Evaluate the probability of the calculated test or sample statistic to make an informed decision regarding rejection of the H_o or fail to reject the H_o

Keywords and Definitions

Inferential statistics	Statistical testing procedures that allow for generalizability of data. They infer from a sample to a population. Through hypothesis testing procedures, these statistics allow to test relationships with the highest level of measurement of variables-interval-ratio continuous or its equivalent (i.e., scaling, or dummy coding).
Sample	A subset of a population.
Population	All the individuals we want to generalize about from a subset of the sample.
Parameter	The fixed value of the population.
Statistic	The calculations we derived from our samples.
Random sampling or Equal Probability Selection Methods (EPSeM)	A probabilistic method of achieving a sample for quantitative research and/or statistics. Everyone has an equal chance of being selected.
Simple random sampling	This sampling method is by far the most common and ensures that everyone in the population has an equal chance of being selected.
Systematic sampling	A form of simple random sampling in which individuals or units are selected from sampling frame or list and a sampling interval is calculated by using the following formula: $Sampling\,interval = \frac{population}{sample}$
Stratified sampling	This randomizes population units and individuals by certain strata or categories, like gender, race, rural, urban, and then randomly selects from those strata or categories.
Multi-stage cluster sampling	This is used when a sampling frame or list is not available or too costly or untimely to produce. When population lists are hard to achieve this sampling is used. This type of sampling utilizes many stages or clusters of sampling, and often covers wide geographic areas in which units or individuals are drawn from.

Convenience samplingmethods	In this method of sampling, participants are selected based on easy recruitment and accessibility. There are no huge costs associated with this. The sample is selected because of comfort, cost, accessibility, and ease.
Sampling bias and error	This occurs when participants are not pooled from a randomized sample and there is no equal representation of selection; as well sample size is compromised.
Population parameters	True parameters or characteristics of the population.
Sample distribution	The sample characteristics.
The sampling distribution	Distribution of all sample outcomes, this contains the sample statistics from the sample. It makes inferences possible.
Confidence levels	These are set to determine the researcher's 'willingness to be wrong'. Confidence levels range from 90%, 95% and 99%, with corresponding alphas of 0.10, 0.05, and 0.001.
Statistically significant	This is determined through our sample statistic and occurs when the null hypothesis is rejected and $p < .05$. The opposite scenario is not statistically significant or when we fail to reject the null hypothesis. Every statistical test has their own statistic.
Type 1 (alpha error)	Rejection of the null hypothesis when the null is true; False-Positive – you decide treatment works when it doesn't really.
Type 2 (beta error)	Failure to reject the null hypothesis when the null is false; you decide treatment doesn't work when really it does.

Test Your Knowledge

1   _____ statistics is a branch of statistics that allows one to generalize from a sample to a population.

    a   Descriptive statistics
    b   Inferential statistics
    c   Descriptive and inferential statistics
    d   EPSeM (Equal Probability Selection Methods or Random Sampling)
    e   Central limit theorem

2   What constitutes an inferential statistical test?

    a   Zero-order cross tabs
    b   Elaborated cross tabs
    c   Frequencies
    d   t-test, ANOVA, Two-way factorial ANOVA, ANCOVA, MANOVA
    e   Univariate statistics

3   _____is determined through our sample statistic and occurs when the null hypothesis is rejected and $p < 0.05$. The opposite scenario is not statistically significant or when we fail to reject the null hypothesis. Every statistical test has their own statistic.

    a   Statistical sampling
    b   Statistical significance
    c   Statistical population parameters
    d   Confidence levels
    e   Sample distribution

4 A researcher's willingness to be wrong is referred to as:

   a   Statistical significance
   b   Confidence level
   c   Type 1 error
   d   Type 2 error
   e   Sampling bias

5 Karen engaged in *systematic sampling*. The population was 100 and the sample was 20. What is the kth interval she should calculate to collect her sample:

   a   Every 2nd person
   b   Every 5th person
   c   Every 20th person
   d   Every other person
   e   Every random person

6 Convenience sampling is a type of random sampling

   a   True
   b   False

7 When there is statistical significance then:

   a $p < 0.05$
   b $p > 0.05$
   c   The null hypothesis is rejected
   d   Fail to reject the null
   e   a and c

8 The sample statistic determines rejection of the null hypothesis.

   a   True
   b   False

9 The goal of inferential statistics is that:

   a   The sampling statistic equate the population parameters
   b   The population parameters equate the sampling statistic
   c   To be able to make inferences from a sample to a population
   d   To report trends and patterns of the data
   e   a and c

10 Inferential statistics are more advanced statistical techniques based on properties of the normal curve.

   a   True
   b   False

6   Bivariate Hypothesis Testing Using an Independent Samples Comparison of Means t-test Using a Dichotomous Independent Variable

Going beyond descriptive trends and patterns of data is important and relevant in data analyses. Sometimes, we do not want to simply explore and describe, we want to answer a specific research question(s) and validate our research hypothesis and confirm our educated guesses are indeed supported by statistical hypothesis testing and sample statistics that align with properties of the normal curve such that generalizability of data is warranted and specific conclusions from a sample to a population can be accurately made.

Hypothesis testing helps to assess, evaluate, and dissect a research problem about the social world. For example, a university evaluates the effectiveness of a hybrid teaching model versus traditional in-person learning models and compares various groups by comparing average differences or an organization may want to assess gender differences and pay scale or to examine weight loss and pair it with self-esteem scales. The possibilities are much larger at the inference-based level. There are more unique ways to analyze data and get an adequate finding(s) that eventually is generalizable, if done correctly and in a meaningful way.

Statistical hypothesis testing in inferential statistics is a method that evaluates various hypotheses about population parameters based on specific sample statistics. In other words, hypothesis testing is heavily invested in making use of a subset of a population, the sample, to formulate conclusions about any population to the best of its abilities. This testing is most often bivariate in nature and involves two variables in which most often the IV are nominal and ordinal and the DV, almost always, is interval-ratio continuous. These hypotheses, in this portion of the book, are represented in terms of 'average' or 'mean' differences. They investigate the most informative and advanced measure of central tendency, the arithmetic average or mean.

This testing requires several assumptions to be met prior to running them and achieving an accurate result. These assumptions include considerations of data, level of measurement, the method of sampling (i.e., random), the shape of the distribution, equal or not equal variances, sample size, confidence level and selected sample statistic based on the test being run.

Assumptions	Explanation
1. Data or level of measurement of variables	The data or level of measurement assumption clearly outlines the types of variables an independent samples t-test is bound to. In this an IV that is dichotomous is key. One that has two response attributes, like men and women, and a DV that is interval-ratio continuous or scaled
2. Sampling method and sample size	Random sampling and large sample

DOI: 10.4324/9781003215691-8

Assumptions	Explanation
3. Independence of observations	Data observations must be independent of each other; no influence by other observations
4. Shape of the distribution	Normally distributed data
5. Homogeneity of variances	Equal or not equal variances assumed, where a preference is given to 'equal' variances. If the F-statistic is significant we report not equal variances; if the F-statistic is not significant we report unequal or not equal variances; equal variances is preferred
6.Confidence level	The amount of risk you are willing to take: 95%, 0.05 alpha
7. Sample statistic	t-statistic or t-value denoted as t: rejects the H_o or fails to reject it
8. Outliers	Minimize outliers or extreme scores

Activity Alert

Carefully, list all assumptions associated with the independent samples t-test.
What are the variable and level of measurement requirement for the independent samples t-test?
What measure of central tendency is key to independent samples t-test?

Hypothesis Testing for Comparison of Means Tests: Independent Samples t-test

Here we begin the inferential statistics journey with our most elementary test of hypothesis testing-the *independent samples t-test*. The t-test family is not very extravagant or powerful and focuses on examining significant mean or average differences amongst two independent groups and two groups only. There are many types of t-test tests, like one sample t-test and paired samples t-test, this chapter focuses on the most common t-test. The most popular t-test used in the social sciences is the independent samples t-test or two-sampled t-test and paired samples t-test (i.e., compare means for the same group at different time intervals). This chapter focuses on the former.

The independent samples t-test is an inferential based parametric test[1] that investigates 'average' or 'mean' differences amongst two groups or samples. The t-test can only test for two groups and not more than that. It is a bivariate test that has a categorical or discrete IV that is dichotomous in nature and a DV that is interval-ratio continuous. For example, an IV with two response attributes, like:

What is your sex?	Training on diversity, inclusion, equity is important in organizations
[0] male	[0] no
[1] female	[1] yes

and a DV that is interval-ratio continuous or its equivalent. An independent samples t-test is an inference based statistical test that is utilized when interested in comparing average or mean differences between two *different* groups and only two groups. This is used in many disciplines across the board and is a useful and popular technique that provides important information about a hypothetical relationship. For example, a one might be interested in 'average differences' in salary and faculty rank (Assistant/Associate Professor vs. Full Professor) or 'mean differences' in intimate partner violence occurrences and family ties in host country or 'average differences' in # of times arrested and gender. All these examples have one common factor: they work with a dichotomous or two response attribute variable for the IV and interval-ratio continuous variable as the DV. It is important to identify the IV and DV and understand the relationship being conceptualized and tested. In quantitative research, research that investigates 'gender differences' adopts to the t-test model in which means are compared across two groups and conclusion made about various dependent variables. The average differences are assessed for statistical significance.

Independent samples t-test can be one-tailed or two-tailed. In a one tailed test, the research hypothesis assumes direction; it specifies that a population mean is either less than (<) or greater than (>) some specified value. There is an increase or decrease in average differences. In a two-tailed test. The research hypothesis specifies that the population mean is not equal to some specified value, there simply is an average or mean 'difference'. For example, a one-tailed test reads like this: Average GPA of Criminology students are lower than the average GPA of all students. A two-tailed test is non-directional. Here, direction of that difference is not specified. The research hypothesis specifies that the population mean is not equal or that there is a 'difference' in mean values. For example, Average GPA of students of Criminology students are different than the average GPA of all students.

Activity Alert

Try to identify the IV and DV in the above examples; also try to create your own examples and discuss. Are your examples one-tailed and two-tailed?
What is the difference between one and two-tailed tests?

The core objective of the independent samples t-test is to compare two sample means from non-related groups. The key question we ask: are the samples different from each other? or are there average or mean differences between the groups in question. In an independent samples t-test you are most often assessing the following scenarios:

- Statistical differences between the means of two groups
- Statistical differences between the means of two interventions
- Statistical differences between the means of two change scores

As stated previously, the main goal in this hypothesis testing technique is to ascertain the 'average or mean differences' and if the calculated mean difference amongst the groups is statistically significant or not. Is there enough evidence to suggest a significant enough difference? There are three essential components of any independent samples t-test.

What variables do you need to run a t-test?

- Independent variable: Nominal or ordinal categorical or discrete variable with only two response attributes or
- Dependent variable: Interval-ratio continuous or scaled.

The research question and the null and research hypothesis is written in specific terms. Let's see the example below:

Research Question: To examine mean or average differences by gender and times being convicted?

Null and research hypotheses:

a   Null hypothesis: There are no statistically significant average differences between the two groups. Example: There are no statistically significant average differences between men and women on number of times being convicted. Both groups have similar or equal means.

b   Research or Alternate hypothesis: There are statistically significant average differences between the two groups. Example: There are no statistically significant average differences between men and women on number of times being convicted. Both groups vary and have different means or not equal.

Please note: The F-statistic speaks to the Homogeneity of Variance assumption. Sometimes there is confusion about this statistic. This statistic assesses equal or not equal variances and that is all-it is examining an assumption of variance.

Homogeneity of Variance Assumption Rules, Levene's Test for Equality of Variances

F-statistic	*F-statistic*
If the probability of the calculated F-statistic is less than 0.05 and statistically significant, then not equal variances are assumed	If the probability of the calculated F-statistic is less than 0.05 and statistically significant, then equal variances are assumed

The sample statistic here is the t-statistic or t-value. This statistic rejects or fails to reject the null hypothesis. This is the statistic that speaks to the research hypothesis. The sample statistic should usually fall in the realm of the null hypothesis being rejected.

Sample Statistic, t-value Interpretation

t-value	*t-value*
If the probability of the calculated t-value is less than 0.05 and statistically significant, suggesting that there are significant mean differences; the null hypothesis is rejected.	If the probability of the calculated t-value is greater than 0.05 and statistically not significant, suggesting that there are no significant mean differences; fail to reject the null hypothesis.

In my years of teaching, I have always noticed that students falter on the F-statistic and t-value. Please recognize the difference between them. The former is testing an assumption of variances, and the latter is testing the research hypothesis. This is a very important difference to recognize and understand.

Effect Size and the Independent Samples t-test

The final statistic that deserves some notice in the independent samples t-test is effect size. Effect size is basically a measure that tells us how small, medium, or large an effect may be. It basically informs us how much variability lies between the two averages by examining standard deviation. For example, while the t-value in an independent samples t-test demonstrates that there is a difference, it does reflect the effect of that difference and the size. A large effect size indicates that the difference is relevant, important, and has meaningful significance. Small effect sizes indicate that there is not much relevance to the calculated difference and medium effect size is good. In an independent samples t-test, Cohen's d is the statistic used to compute effect size of the difference between two groups. Cohen's d establishes the size or impact of the effect and eventually how meaningful the relationship or mean difference amongst variables is.

Effect Size	Cohen's d
Small	0.2
Medium	0.5
Large	0.8 or greater

It always good to mention Cohen's d when analyzing an independent samples t-test because it really provides perspective of the significant differences. SPSS does calculate this.

Key Statistics to Report

1   Descriptive statistics: Report basic descriptives for each group and include the sample size (N), averages or means, mean difference, and standard deviation: provide trends and patterns of data; this provides readers with a quick overview of the key group characteristics as it relates to the DV or social phenomenon.

2   Levene's Test of Homogeneity of Variances: Report whether equal or not equal variances are assumed; equal ($p > 0.05$) or not equal variances ($p < 0.05$): Tests the assumption of equal or not equal variances, remember you can never have two nots together (i.e., not significant, means equal variances; significant means not equal variances)

3   t-value: Report the calculated test statistic and evaluate statistical significance to reject or fail to reject the null hypothesis; the t-value is the difference between means relative to the variability within groups. Degrees of freedom can be reported but is not mandatory.

4   Cohen's d (effect size): magnitude of effect size and whether a meaningful difference is apparent; this quantifies effect size and difference of two groups.

5   Always tell the reader, what the confidence level and alpha is set to. For example, is it at 95%, alpha of 0.05 or 99%, alpha of 0.001?

The independent samples t-test, while a valuable data analytical test, does come with some noteworthy limitations. First, it is sensitive to outliers, especially with small samples; second, departures from normal distributions can impact the quality and accuracy of the test; third, the validity of the t-test can be compromised if equal variances are not present; fourth, an independent samples t-test is limited to only two groups at a time. Data must always be modified if there are more than two groups. In this way, certain specifics or details of variable are lost. Also, you need a reasonable sample size to make inferences, as well as normally distributed properties. For example, a race variable that his multiple response attributes, needs to be recoded to [0] whites and [1] minorities. However, to analyze more than two groups of racial background, other statistical tests belonging to the ANOVA family have been devised to take on the multiple groups. The next chapters examine this family in depth and advantages of using of ANOVA in your statistical journey and pathways.

$$t = \frac{\bar{x}_1 - \bar{x}_2}{s_p \cdot \sqrt{\frac{1}{n_1} + \frac{1}{n_2}}}$$

$$s_p = \sqrt{\frac{(n_1 - 1)s_1^2 + (n_2 - 1)s_2^2}{n_1 + n_2 - 2}}$$

Independent Samples t-test in SPSS

1 Click on Analyze > Compare Means and Proportions > Independent Samples t-test
2 Move the variables of interest into the Test Variable Column (DV) and Grouping Variable (IV); Define grouping variable codes, as 0 or 1 or 1 or 2
3 Check off Estimate Effect Size
4 Click OK to generate an independent samples t-test

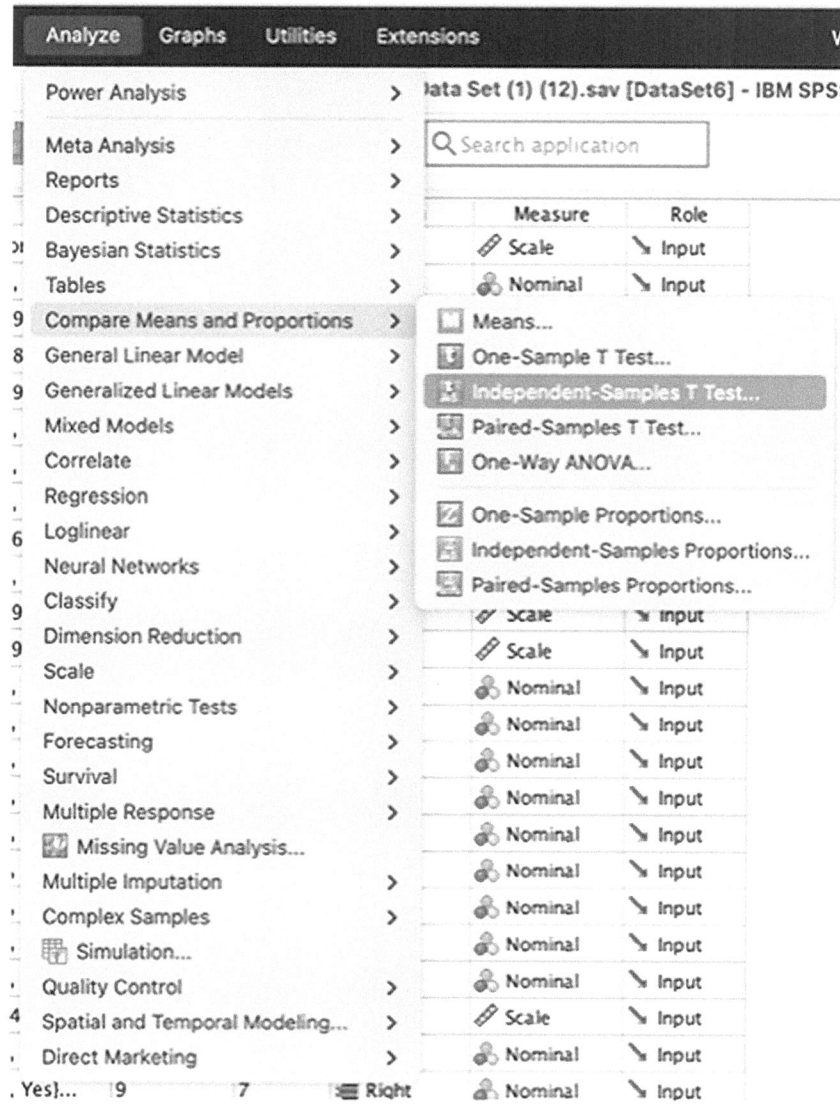

Figure 6.1a SPSS Command for an Independent Samples t-test

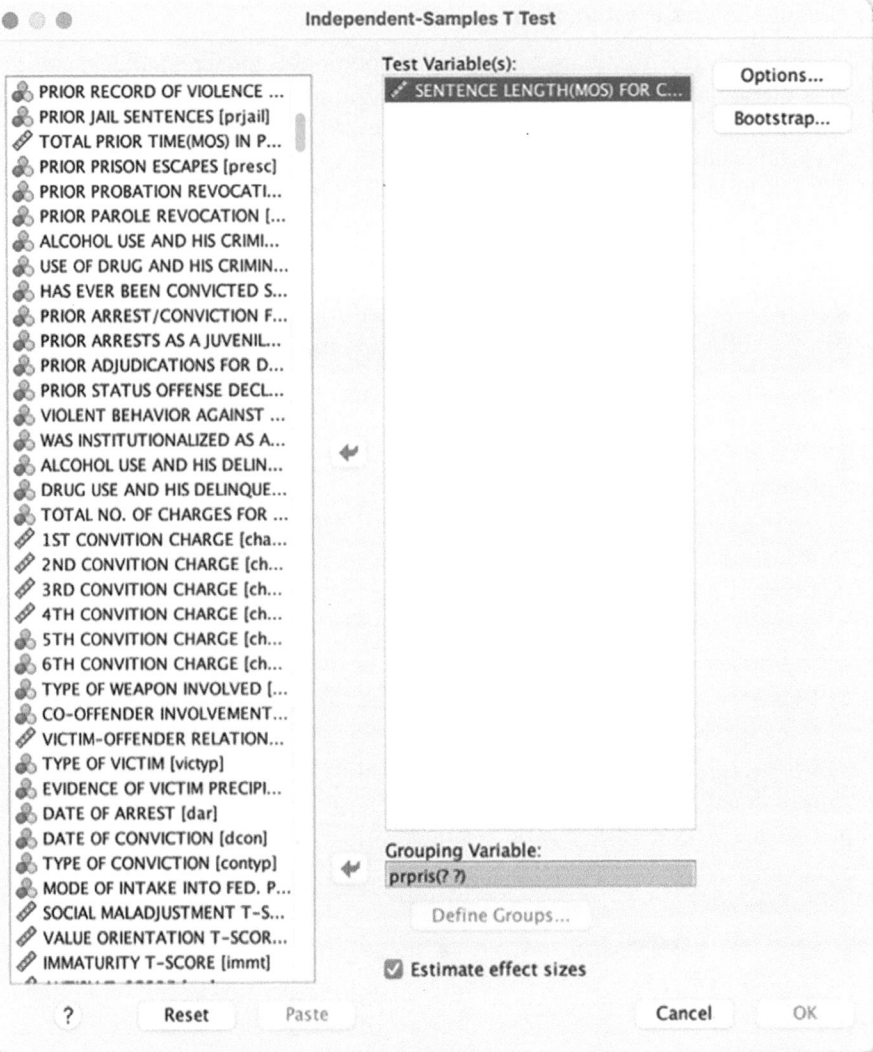

Figure 6.1b Add the Test Variable, DV to the Test Variable Box and Add the Grouping Variable, IV to the Grouping Variable Box and Ensure Groups are Defined; Estimate of Effect Size (Cohen's d) Should be Checked-off

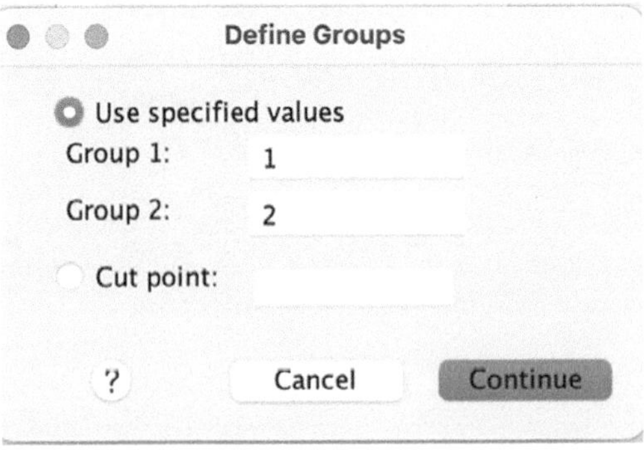

Figure 6.1c Ensure Groups are Defined, as 1 and 2, or 0 and 1

Group Statistics

	INMATE SERVED A PRIOR PRISON SENTENCE	N	Mean	Std. Deviation	Std. Error Mean
SENTENCE LENGTH(MOS) FOR CURRENT OFFENSE	Yes	159	111.95	103.844	8.235
	No	187	83.65	155.293	11.356

Independent Samples Test

		Levene's Test for Equality of Variances		t-test for Equality of Means							
						Significance		Mean Difference	Std. Error Difference	95% Confidence Interval of the Difference	
		F	Sig.	t	df	One-Sided p	Two-Sided p			Lower	Upper
SENTENCE LENGTH(MOS) FOR CURRENT OFFENSE	Equal variances assumed	.073	.787	1.956	344	.026	.051	28.303	14.470	-.158	56.763
	Equal variances not assumed			2.018	326.706	.022	.044	28.303	14.028	.706	55.899

Independent Samples Effect Sizes

		Standardizer[a]	Point Estimate	95% Confidence Interval	
				Lower	Upper
SENTENCE LENGTH(MOS) FOR CURRENT OFFENSE	Cohen's d	134.135	.211	-.001	.423
	Hedges' correction	134.429	.211	-.001	.422
	Glass's delta	155.293	.182	-.030	.394

a. The denominator used in estimating the effect sizes.
   Cohen's d uses the pooled standard deviation.
   Hedges' correction uses the pooled standard deviation, plus a correction factor.
   Glass's delta uses the sample standard deviation of the control (i.e., the second) group.

Figure SPSS Output #6 Comparison of Means-Independent Samples t-Test

SPSS Example: Technical and Substantive Interpretation

Comparison of Means Hypothesis Testing: Independent Samples t-test

Research Question: Are there average or mean differences in sentence length for current offence and inmate having served a prior prison sentence?

Null Hypothesis: There are no statistically significant average or mean differences seen in sentence length for current offence and inmate having served a prior prison sentence. Those prisoners with a prior sentence will not have a greater sentence length for current offence. The means are equal.

Research Hypothesis: There are statistically significant average or mean differences seen in sentence length for current offence and inmate having served a prior prison sentence. Those prisoners with a prior sentence will have a greater sentence length for current offence. The means are different.

Technical interpretation

Group statistics: The sample size for the dichotomous independent or grouping variable, inmate served a prior sentence or not is 159 (46%) and 187 (54%) respectively out of 346 (N). The measure of central tendency, mean, indicates that on average, sentence length in months for prisoners who served a prior sentence was about 112 months and those that did, on average had prison sentence lengths for about 84 months, with standard deviation s of 103.84 and 155.29 correspondingly. There is a mean or average difference of approximately 28 months between those that served prior prison sentence and sentence length.

Levene's Test of for Equality of Variances: The Levene's test for Equality of Variances suggest a not significant outcome (F = 0.073, $p > 0.05$) and therefore, equal variances are assumed. Therefore, the t-value that will be evaluated is for 'equal' variances.

Test Statistic and Cohen's d: The t-value (t = 0.026, $p < 0.05$) is statistically significant and the null hypothesis stating no average differences or means are equal is rejected. Cohen d = 0.211 is a 'small' effect size, indicating the difference between these two groups is weak or small. There may be less variability amongst the means of both groups.

Substantive interpretation

Thus, there were clearly more prisoners in the sample that did not serve a prior sentence, compared to those that did. However, average differences between the two groups clearly indicated that those prisoners that served a prior sentence had greater sentence lengths, compared to those that did not, but was deemed weak. There are average differences seen in the two groups. The sample sizes were different in terms of representation, and this should be taken into consideration when assessing the t-test. Resume with caution as inferences are made from a sample to a population.

Final Thoughts

Bivariate hypothesis testing of comparison of means or averages comes in all shapes, sizes, and forms and begins the discussion surrounding the other major branch of statistics, inferential statistics. This chapter dealt with one of the simplest forms of comparison of means tests, the independent samples t-test. This type of t-test analysis is dedicated to a dichotomous IV and an interval-ratio continuous variable and seeks to examine statistically significant ($p < 0.05$) average or mean differences between two groups and two groups only. Any data modification here is done to transform variables into dichotomies. Independent samples t-test are parametric tests that are heavily assumption based. The sample statistic for the independent samples t-test is the t-value and the estimate of effect size, Cohen's d is a measure of magnitude and if the average difference is large enough to be meaningful. Data or level of measurement of variables, sampling method and sample size, Independence of observations, shape of the distribution, homogeneity of variances, confidence level, sample statistic, and outliers are the assumptions that must be met prior to running a t-test to get the real picture of data. Overall, the independent samples t-test is simply, applicable to small sample sizes, versatile, robust to deviations of normality, and easy to interpret. The interpretation of outputs is not cumbersome or overly complex. Thus, storytelling of numbers here is relaxed and quite succinct.

Keywords and Definitions

Statistical hypothesis testing	In inferential statistics this is a method that evaluates various hypotheses about population parameters based on specific sample statistics.
Comparison of means	The examination of 'average' or 'mean' differences in a sample, with the hope of generalizing findings from a sample to a population.
Independent samples t-test	An inference-based test that examines a dichotomous IV and an interval-ratio DV.
Homogeneity of variance	This is an assumption of the independent samples t-test. It evaluates whether variances are equal or not. It does not reject the null hypothesis.
t-value	This is the sample statistic here is the t-statistic or t-value. This statistic rejects or fails to reject the null hypothesis. This is the statistic that speaks to the research hypothesis.
Cohen's d	This establishes the size or impact of the effect and eventually how meaningful the relationship or mean difference amongst variables. It can be large, medium, or small.

Test Your Knowledge

1 _____testing in inferential statistics is a method that evaluates various hypotheses about population parameters based on specific sample statistics.

   a   Statistical hypothesis testing
   b   ANOVA
   c   Levene's Test for equality of variances
   d   Independent samples t-test
   e   MANOVA

2 _____establishes the size or impact of the effect and eventually how meaningful the relationship or mean difference amongst variables. It can be large, medium, or small.

    a   Cohen's d
    b   Cohen's A
    c Eta^2
    d   Adjusted R^2
    e   None of the above

3 A dichotomous IV and interval-ratio DV is used on two different groups when running which test?

    a   ANOVA
    b   Paired samples t-test
    c   Independent samples t-test
    d   Regression
    e   Univariate statistics

4 Equal variances are assumed when:

    a   F-stat is less than 0.05
    b   t-value is less than 0.05
    c   F-stat is greater than 0.05
    d   t-value is greater than 0.05
    e   Equal variances do not matter

5 The independent samples t-test has the following assumptions:

    a   Data, sampling, normal distribution, equality of variances, small sample size
    b   Data, sampling, normal distribution, equality of variance, large sample size
    c   It has no assumptions
    d   Only data being dichotomous is the only assumption
    e   None of the above

6 A researcher running an independent samples t test can have varying confidence levels and select amongst 90%, 95% and 99%.

    a   True
    b   False

7 Properties of ___ curve is essential for parametric tests like the t-test.

    a   Chi-Square
    b   Normal
    c   Traditional
    d   t-value
    e   Pooled variances

8 A parametric test is distribution free and has no essential guidelines. A non-parametric test follows properties of the normal distribution.

    a   True
    b   False

9   The key statistics in an independent samples t-test are:

a   Descriptives, sample size, averages, standard deviation, F-stat, t-value, and significance, as well as Cohen's d
b   Inferential statistics, sample size, averages, standard deviation, F-stat, t-value and significance, as well as Cohen's d
c   There are no statistics to report
d   Dichotomous IV and interval-ratio DV
e   None of the above

10   One of the biggest limitations of t-test are:

a   Assumptions
b   Can only take on a dichotomous IV by assessing two groups
c   Very complex
d   Only assesses averages
e   a and b

Note

1 Parametric tests follow properties of the normal distribution and work with interval-ratio data, the highest level of measurement.

7 Bivariate Hypothesis Testing Using Comparison of Means using More than Two Groups, One-Way ANOVA

The ANOVA, analysis of variance family, is a sizeable group of statistical tests. It is referred to as 'analysis of variance' because it diligently captures and compares the variance or variability of scores between various groups with the variability within the groups. This test allows for efficient comparisons of average differences across multiple groups. Rather than running multiple t-tests for every single combination of variables to assess significant average differences and increase Type 1 errors, ANOVA allows to simultaneously do this in one test.

There are many types of analysis that fall into this grouping, like One-way ANOVA, Two-way factorial ANOVA, ANCOVA, MANOVA and MANCOVA. The list is elaborate, and the ANOVA family spectrum is vast. All these tests, in their own manner, compare means using hypothesis testing procedures. The analysis is heavily linked to 'average differences', like the t-test. In the social world, there are many non-dichotomous instances we come across, like marital status or even gender nowadays has multiple response attributes. Our world is preoccupied with varying choices and possibilities. Not everything can be measured or assessed in a binary method. Certain aspects of social life require non-binary variables to be assessed at best. The ANOVA family is heavily linked to the discipline of psychology. Psychology, due to its experimental nature, has always been a frontrunner in the use of ANOVA but other disciplines, like sociology and criminology, are making use of this statistical test. Sociologists' and criminologists' use of the ANOVA statistical tests are trending. One-way ANOVA is a versatile statistical test that can be applied in many unique ways to various scenarios.

Hypothesis Testing for Comparison of Mean Tests: One-Way ANOVA

This chapter focuses specifically on the One-way ANOVA, or analysis of variance, parametric test – an alternative to the independent samples t-test. Like t-test, the One-way ANOVA is also identified as a comparison of mean test hypothesis, testing bivariate statistical analysis examining an IV and DV. Here, the IV is referred to as a factor or main effect. When speaking of IVs in ANOVA, the preference is to refer to the IVs as factors or main effect(s). You should make this a habit. The main effect examines the *independence* of effects of each factor as it relates to the DV or outcome, regardless of other parameters or factors. The biggest difference is the amount of groups t-test and One-way ANOVA work with. One-way ANOVA is an extension of the independent samples t-test which takes on more than two groups' means (i.e., non-dichotomous variables). A One-way ANOVA design focuses on a single independent variable and single dependent variable and if most often referred to as a univariate design. There are no control variables or covariate variables in this design.

DOI: 10.4324/9781003215691-9

For example, a variable like:

How much do you agree with the following statement: The COVID-19 pandemic has created feelings of isolation, especially with elderly population?

[0] strongly agree
[1] agree
[2] disagree
[3] strongly disagree

What is the racial category that *best* represents you?

[0] White
[1] South Asian
[2] Chinese, Japanese, Korean
[3] Black
[4] Filipino
[5] Latin American
[6] Arab
[7] Other

These two examples are clearly non-dichotomous variables, and both have more than two groups to compare across and between groups. Clearly, these multi-level variables have multiple, meaningful groups that are relevant to the research or research questions. The beauty of ANOVA is that it takes on variables in their natural state without the need for recoding or dummy coding. Quantitative researchers get all the attention to detail of each variable and see each group as an opportunity for statistical analysis and discussion. For instance, the race variable in its recoded state would minimize the details of the variable by grouping it into [0] whites [1] minority. In a t-test analysis that would have to be done and it is evident that details of the variable's response attributes are lost. Thus, One-way ANOVA allows us to pursue specified grouping and details. This test also assesses statistically significant 'average' or 'mean' differences for more than two groups. A One-way ANOVA makes use of one independent variable or factor or one main effect that is nominal or ordinal nature and one dependent variable that is interval-ratio continuous or its equivalent. It simply assesses one 'main effect' that has more than two groups – the main effect is the effect that one IV has on the DV, regardless of other factors or variables. This statistical test is slightly more informative than independent samples t-test because it allows us to analyze more than two groups, without any data modification, by keeping the original state of the factor or IV. When three or more groups are tested, statistically significant differences can be analyzed. In a One-way ANOVA, you are most often assessing the following scenarios:

- Statistical differences between the means of two or more groups
- Statistical differences between the means of two or more interventions
- Statistical differences between the means of two or more change scores

Each group mean is compared to the *grand mean*. The grand mean is the baseline average of all data points for all groups or conditions for a specific DV. Basically, the grand mean represents the overall average score. Each group mean is compared to the grand mean as being above or below the grand mean value. This provides important information about each group's average, compared to the grand mean value, and must not be taken lightly. Reporting these highs and lows is critical to the analysis phase.

Like the independent samples t-test, One-way ANOVA has assumptions that must be met for reliable and valid outcomes or results. If assumptions are not met, it must list as a limitation of the test and that findings may not be generalizable from a sample to a population.

What variables do you need to run a One-way ANOVA?

- IV or factor: One nominal or ordinal categorical or discrete variable with more than two response attributes or
- Dependent variable: One interval-ratio continuous or scaled.

Assumptions	Explanation
1. Data or level of measurement of variables	The data or level of measurement assumption clearly outlines the types of variables a One-way ANOVA is bound to. In this an IV or factor that is nominal or ordinal categorical or discrete is key and a DV that is interval-ratio continuous or scaled
2. Sampling method and sample size	Random sampling and large sample
3. Independence of observations	Data observations must be independent of each other; no influence by other observations
4. Shape of the distribution	Normally distributed data
5. Homogeneity of variances	Equal or not equal variances assumed, where a preference is given to 'equal' variances. If the F-statistic is significant we report not equal variances; if the F-statistic is not significant we report unequal or not equal variances
6. Confidence level	The amount of risk you are willing to take: 95%, 0.05 alpha
7. Sample statistic	F-statistic or F-ratio: rejects the H_o or fails to reject it
8. Outliers	Minimize outliers or extreme scores

Activity Alert

Carefully list all assumptions associated with the One-way ANOVA.
Should variances be equal or not for the homogeneity of variance test?
How does the One-way ANOVA family differ from the t-test family?

The research question and the null and research hypothesis are written in specific terms. Let's see the example below:

Research Question: To examine mean or average differences by race and times being convicted?

Null and research hypotheses:

a   Null hypothesis: There are no statistically significant average differences in the main effect of all groups.

Example: There are no statistically significant average differences in the main effect of race and times being convicted. All group means are equal (i.e., same)
$\mu1 = \mu2 = \mu3 = \mu4$

b *Research or alternate hypothesis*: There are statistically significant average differences in the main effect of all groups.Example: There are statistically significant
average differences in the main effect of race and times being convicted. All
group means are not equal (i.e., different); at least two differ.
$\mu1 \neq \mu2 \neq \mu3 \neq \mu4$

Please note: Similarly, to the independent samples t-test, the F-statistic speaks to the
Homogeneity of Variance assumption. Sometimes there is confusion about this statistic. This statistic assesses equal or not equal variances and that is all-it is examining an
assumption of variance. In ANOVA, variances must be equal. If they are not equal,
the assumption has been violated and a correction needs to be put in place and this
must be discussed as a limitation of the test. If the Levene's test does not maintain
'equal' variances and the variances differ, then F_{welch} is run, as the correction factor
sample statistic.

Homogeneity of Variance Assumption Rules, Levene's Test for Equality of Variances

F-statistic	*F-statistic*
If the probability of the calculated F-statistic is less than 0.05 and statistically significant, then not equal variances are assumed	If the probability of the calculated F-statistic is less than 0.05 and statistically significant, then equal variances are assumed

The sample statistic here is the t-statistic or t-value. This statistic rejects or fails to reject
the null hypothesis. This is the statistic that speaks to the research hypothesis. The sample
statistic should usually fall in the realm of the null hypothesis being rejected.

Sample Statistic, F-value Interpretation when Equal Variances are Retained:

Original F-value or F-ratio	*Original F-value or F-ratio*
If the probability of the calculated F-statistic is less than 0.05 and statistically significant, suggesting that there are significant mean differences; the null hypothesis is rejected.	If the probability of the calculated F-statistic is greater than 0.05 and statistically not significant, suggesting that there are no significant mean differences; fail to reject the null hypothesis.

Sample Statistic, F-value Interpretation when Not Equal Variances are Retained:

Welch's F-statistic (F_{welch})	*Welch's F-statistic (F_{welch})*
If the probability of the calculated F_{welch} statistic is less than 0.05 and statistically significant, suggesting that there are significant mean differences; the null hypothesis is rejected.	If the probability of the calculated F_{welch} statistic is greater than 0.05 and statistically not significant, suggesting that there are no significant mean differences; fail to reject the null hypothesis.

The sample statistic for One-way ANOVA is the F-statistic when variances are
equal; this analyzes mean differences among the multiple groups in a sample. Again,
the sample statistic informs us that there are statistically significant mean differences
among the groups or response attributes or categories. If not, equal variances are
present, the Welch's F or F_{welch} is the designated statistic to use. This statistic is quite
useful when group variances differ, as well as sample size.

One-Way ANOVA and Post-hoc Testing

In either scenario, F-statistic or F_{welch} these sample statistics, only highlight an average or mean difference, but does not inform us of which groups significantly differ or not. To establish precision in mean differences, a post-hoc test needs to be run to establish exactly where the group differences lie. This is important to know. Knowing that there are mean differences is great, but to complete the analysis reporting of which groups significantly differ is equally important to this type of analysis. There are many post-hoc testing procedures based on the research design and nature of data, like Tukey, Scheffé, Dunnett's, and Bonferroni, that the ANOVA family offers. Bonferroni post-hoc test is the most robust one and popular one that is generally used. Post-hoc testing is an integral part of the One-Way ANOVA test and must be discussed for a complete analysis of results.

Effect Size and One-Way ANOVA

Like independent samples t-test, One-way ANOVA also provides an estimate of the effect size of the relationship being tested. An eta^2 is calculated to explain how much variance in the DV is being explained by the main effect or factor or IV. The closer to 100% explanation, the better the explanation. No research finding will give 100% explanation, but the closer the value is to 100 the better the explained variance. These can be explained by weak, moderate, or strong effects. For example, if an eta^2 is calculated to be 0.313, 0.313 x 100 would indicate that 31.3% of the variance in #of times convicted is explained by the main effect of race. A somewhat moderate effect.

Key Differences Between an Independent Samples t-test and One-way ANOVA

Characteristic	Independent samples t-test	One-way ANOVA
Level of measurement & Types of variables	One dichotomous IV and one interval-ratio continuous DV	One non-dichotomous or more than two groups IV and one interval-ratio continuous DV
Statistical analysis	Compares average differences between *only* two groups	Compares average differences between more than two groups
Post-hoc tests	No post-hoc testing	Engages in post-hoc testing
Key assumption	Normal distribution of the DV	Assumes equal variances across groups and normally distributed residuals
Statistical test	Provides a t-statistic to reject the null hypothesis	Provides an F-statistic to reject the null hypothesis

Key Statistics to Report

1   Descriptive statistics: Report basic descriptive statistics, sample size (N), averages, grand mean, standard deviation, minimum and maximum values; provide trends and patterns of data; this provides readers with a quick overview of each group characteristics as it relates to the DV or social phenomenon.
2   Levene's Test for Equality of Variances: Report whether equal or not equal variances are assumed; equal ($p > 0.05$) or not equal variances ($p < 0.05$): Tests the

assumption of equal or not equal variances, remember you can never have two nots together (i.e., not significant, means equal variances; significant means not equal variances); non-violation of the assumption yields equal variances amongst groups.

3   F-value or F-ratio or F-statistic: Report the calculated test or sample statistic that evaluates statistical significance to reject or fail to reject the null hypothesis of overall comparison of means; or the correction factor, F-welch if not equal variances are assumed. This value reports significant mean differences or not. Degrees of freedom can be reported but is not mandatory.

4   Partial eta^2 (effect size): Magnitude of effect size and reports the proportion of total variance attributed to group differences and whether a meaningful difference is apparent.

5   Post-hoc testing: Report specific significant average or mean group differences. Bonferroni is the most robust test. Other post-hoc test may also be run.

6   Always tell the reader, what the confidence level and alpha is set to. For example, is it at 95%, alpha of 0.05 or 99%, alpha of 0.001?

One-way ANOVA, although a useful statistical data analysis tool has some limitations. First, one-way ANOVA is most often limited to cross-sectional data types. Repeated measures or mixed-effect modelling may be more suitable for longitudinal or repeated measures; second, One-way ANOVA is sensitive missing cases and this may introduce bias, thus missing data should be handled carefully and systematically; additionally, this technique is dependent on only one IV or factor or main effect which makes the analysis limiting; furthermore, One-way ANOVA is bound to assumptions, any violations of these may result in unreliable or invalid data. A reasonable sample size is indeed required to make inferences, as well as normally distributed properties and all the other assumptions that go with it to make sound conclusions at the end. Be cautious in your use of One-way ANOVA.

Source of Variation	Sum of Squares (SS)	Degrees of Freedom (d.f.)	Mean Square (MS) (This is SS Divided by d.f.) and is an Estimation of Variance to be Used in F-ratio	F-ratio
Between samples or categories	$n_1(\overline{X}_1 - \overline{\overline{X}})^2 + \cdots + n_k(\overline{X}_k - \overline{\overline{X}})^2$	$(k-1)$	$\dfrac{SS \text{ between}}{(k-1)}$	$\dfrac{MS \text{ between}}{MS \text{ within}}$
Within samples or categories	$\sum(X_{1i} - \overline{X}_1)^2 + \cdots + \sum(X_{ki} - \overline{X}_k)^2$ $i = 1, 2, 3, \ldots$	$(n-k)$	$\dfrac{SS \text{ within}}{(n-k)}$	
Total	$\sum(X_{ij} - \overline{\overline{X}})^2$ $i, j = 1, 2, 3, \ldots$	$(n-1)$		

Figure 7.1 One-Way ANOVA Formula

One-Way ANOVA in SPSS

1   Click on Analyze > Compare Means and Proportions > One-Way ANOVA
2   Move the variables of interest into the Dependent List (DV) and Factor (IV)
3   Click on Post Hoc >Check off Bonferroni
4   Click on Options >Check off Descriptives, Homogeneity of Variance Test, and Welch Test
5   Check off Estimate effect size for overall test
6   OK to generate a One-Way ANOVA

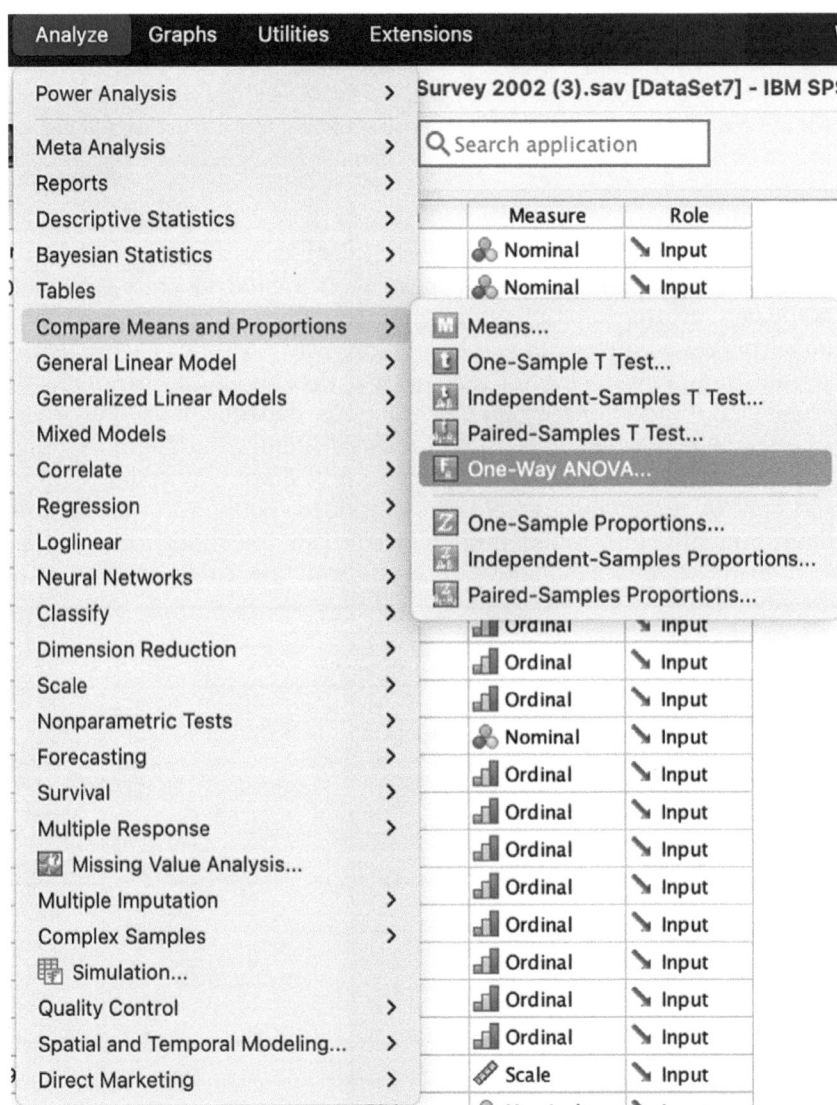

Figure 7.2a SPSS Command for One-Way ANOVA

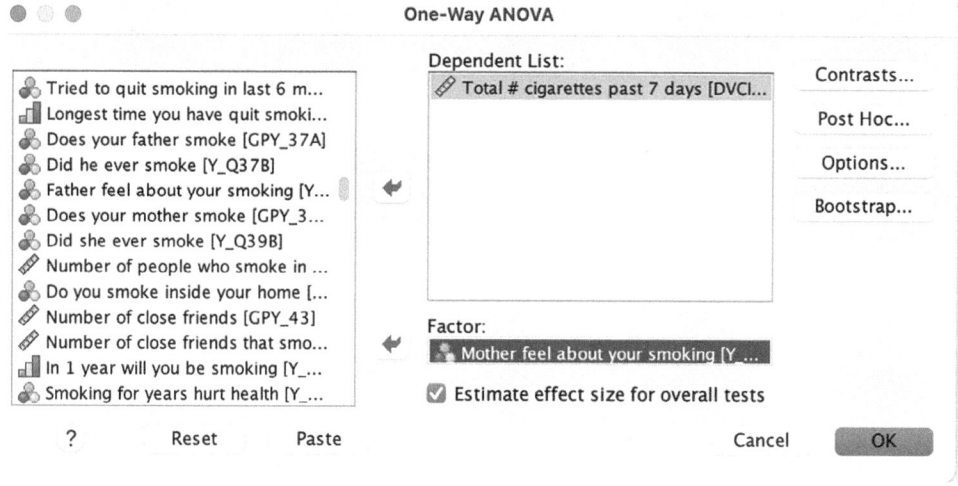

Figure 7.2b Add the DV to the Dependent List Box and the IV to the Factor Box

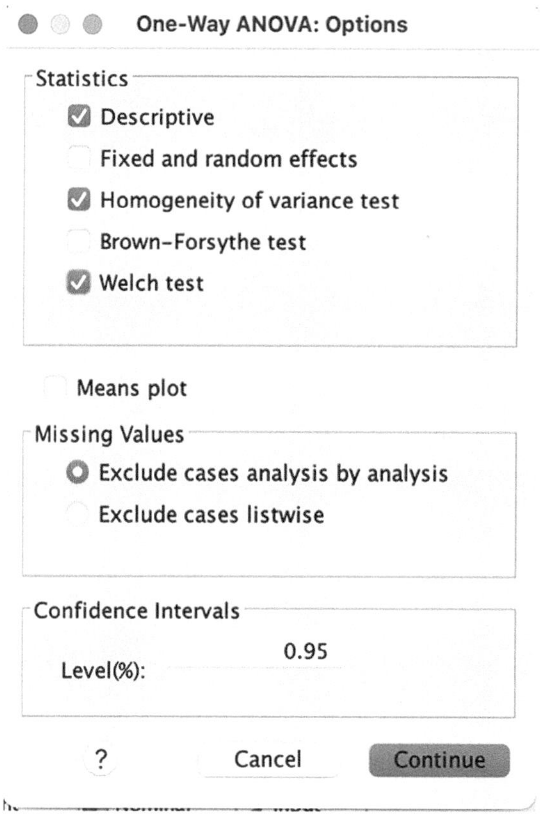

● ◉ ●   **One-Way ANOVA: Options**

Statistics

☑ Descriptive

Fixed and random effects

☑ Homogeneity of variance test

Brown–Forsythe test

☑ Welch test

Means plot

Missing Values

◉ Exclude cases analysis by analysis

Exclude cases listwise

Confidence Intervals

0.95

Level(%):

?   Cancel   Continue

Figure 7.2c Click Options and Ensure Descriptive, Homogeneity of Variance Test and Welch
Test is all Checked-off

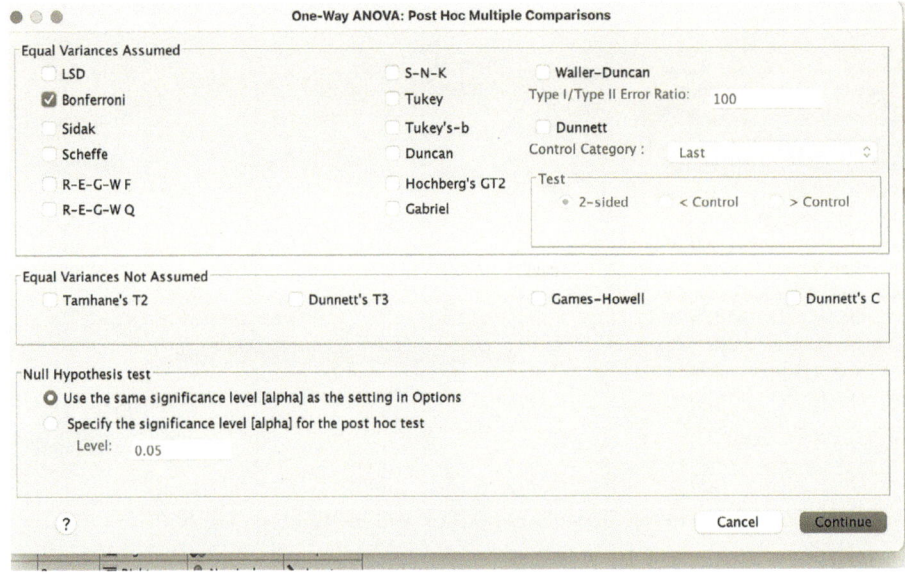

Figure 7.2d Click Post-Hoc and Ensure Bonferroni is Checked-off

Descriptives

Total # cigarettes past 7 days

	N	Mean	Std. Deviation	Std. Error	95% Confidence Interval for Mean Lower Bound	95% Confidence Interval for Mean Upper Bound	Minimum	Maximum	Between-Component Variance
She approves	30	65.83	45.730	8.349	48.76	82.91	6	175	
She doesn't care	110	59.01	41.703	3.976	51.13	66.89	1	190	
She doesn't like it	275	39.30	35.873	2.163	35.04	43.56	0	210	
She doesn't know that I smoke	373	15.26	25.446	1.318	12.67	17.85	0	252	
Total	788	31.68	36.999	1.318	29.10	34.27	0	252	
Model Fixed Effects			32.804	1.169	29.39	33.98			
Random Effects				13.067	-9.90	73.27			461.811

Tests of Homogeneity of Variances

		Levene Statistic	df1	df2	Sig.
Total # cigarettes past 7 days	Based on Mean	35.318	3	784	<.001
	Based on Median	32.717	3	784	<.001
	Based on Median and with adjusted df	32.717	3	773.890	<.001
	Based on trimmed mean	36.225	3	784	<.001

ANOVA

Total # cigarettes past 7 days

	Sum of Squares	df	Mean Square	F	Sig.
Between Groups	233658.960	3	77886.320	72.378	<.001
Within Groups	843669.359	784	1076.109		
Total	1077328.319	787			

Because Variances are not equal, we proceed to F-welch for the correction.

ANOVA Effect Sizes[a]

		Point Estimate	95% Confidence Interval Lower	95% Confidence Interval Upper
Total # cigarettes past 7 days	Eta-squared	.217	.167	.263
	Epsilon-squared	.214	.164	.260
	Omega-squared Fixed-effect	.214	.164	.260
	Omega-squared Random-effect	.083	.061	.105

a. Eta-squared and Epsilon-squared are estimated based on the fixed-effect model.

Figure SPSS Output #7 Comparison of Means – One-Way ANOVA

Robust Tests of Equality of Means

Total # cigarettes past 7 days

	Statistic[a]	df1	df2	Sig.
Welch	64.374	3	114.646	<.001

a. Asymptotically F distributed.

Post Hoc Tests

Multiple Comparisons

Dependent Variable: Total # cigarettes past 7 days
Bonferroni

(I) Mother feel about your smoking	(J) Mother feel about your smoking	Mean Difference (I–J)	Std. Error	Sig.	95% Confidence Interval	
					Lower Bound	Upper Bound
She approves	She doesn't care	6.824	6.757	1.000	-11.05	24.70
	She doesn't like it	26.532[*]	6.307	<.001	9.85	43.21
	She doesn't know that I smoke	50.571[*]	6.225	<.001	34.10	67.04
She doesn't care	She approves	-6.824	6.757	1.000	-24.70	11.05
	She doesn't like it	19.707[*]	3.701	<.001	9.92	29.50
	She doesn't know that I smoke	43.746[*]	3.559	<.001	34.33	53.16
She doesn't like it	She approves	-26.532[*]	6.307	<.001	-43.21	-9.85
	She doesn't care	-19.707[*]	3.701	<.001	-29.50	-9.92
	She doesn't know that I smoke	24.039[*]	2.607	<.001	17.14	30.94
She doesn't know that I smoke	She approves	-50.571[*]	6.225	<.001	-67.04	-34.10
	She doesn't care	-43.746[*]	3.559	<.001	-53.16	-34.33
	She doesn't like it	-24.039[*]	2.607	<.001	-30.94	-17.14

*. The mean difference is significant at the 0.05 level.

SPSS Example: Technical and Substantive Interpretation

Comparison of Means Hypothesis Testing: One-Way ANOVA

Research Question: Are there average or mean differences in total number of cigarettes smoked in past seven days and how your mother feels about your smoking?

Null Hypothesis: There are no statistically significant average or mean differences seen in total number of cigarettes smoked in past seven days and how your mother feels about your smoking? The means are equal.

Research Hypothesis: There are no statistically significant average or mean differences seen in total number of cigarettes smoked in past seven days and how your mother feels about your smoking? The means are equal.

The means are different.

Technical interpretation

Descriptive statistics: The sample size for the non-dichotomous independent or factor or main effect variable, how mother feels about your smoking, indicates that 30 approve, 110 do not care, 275 don't like it, and 373 said their mothers did not know. There were 788 participants in total in which about 47% of mothers were unaware that their child smoked. The total number of cigarettes varied by each group of how mothers feel. For instance, when mothers approved the average number of cigarettes was quite large at about 66, with a standard deviation of 45.73. For those mothers that did not care, the total number of cigarettes was also large at about 59 cigarettes in the past seven days with a standard deviation of 41.70. For those mothers that did not like it, participants smoked, on average, 39 cigarettes were smoked in past seven days with a standard deviation of 35.87. All these groups fell above the grand mean of approximately 32 cigarettes smoked in past seven days. For those mothers that did not know that the participant smokes, on average they smoked the least amount at about 15 cigarettes in the past seven days with a standard deviation of 25.45. This group fell below the grand mean. The minimum and maximum values varied for each group, with the highest maximum number of cigarettes smoked in past seven days amounting to 252 and the lowest 0.

Levene's test for Equality of Variances: Levene's test for Equality of Variances suggests a statistically significant outcome ($F = 35.318$, $p < 0.05$) and therefore, equal variances are not assumed. Because equal variances are not assumed, F_{welch} will be assessed.

Test statistic, Post-hoc test, and Effect size estimate: The F-statistic ($F = 64.37$, $p < 0.05$) is statistically significant suggestive of average group differences between how mother feels about smoking and the number of cigarettes smoked in the past seven days. The null hypothesis is rejected, that means are equal. The sample statistic suggested that there are significant mean differences but did not allude to exactly where the significant differences lie. The Post-hoc test, Bonferroni, revealed that significant average differences ($p < 0.05$) occurred between the following groups:

- She approves – She doesn't like it (26.53) and She doesn't know that I smoke (50.57)
- She doesn't care – She approves (6.82), She doesn't like it (19.71) and She doesn't know that I smoke (43.75)
- She doesn't like it – She doesn't know that I smoke (24.04)

There were only two groups – 'she approves' 'she doesn't care' – which did not result in significant average differences (6.824, $p > 0.05$). The effect size indicated that about 22% of the variance in the number of cigarettes smoked in past seven days is explained by how mother feels about smoking.

Substantive interpretation

In summary, the one-way ANOVA clearly demonstrated that how mother feels about smoking results in varying mean differences in the number of cigarettes smoked in the past seven days. Mother's that approve result in increased smoking, on average, compared to those that don't. There are significant average differences seen in the groups with the greatest average differences between those mothers that approve and those where the mother doesn't know they smoke. Surprisingly, all groups had significant average differences except one, she approves, and she doesn't care. Some notable points to consider when analyzing this one-way ANOVA is that the sample sizes were different in terms of representation. Also, equal variances were not assumed. Resume with caution in making inferences from the sample to the population at hand.

Final Thoughts

The ANOVA, analysis of variance family, is a sizeable group of statistical tests. It is referred to as 'analysis of variance' because it diligently captures and compares the variance or variability of scores between various groups with the variability within the groups. This test allows for efficient comparisons of average differences across multiple groups. Rather than running multiple t-tests for every single combination of variables to assess significant average differences and increase Type 1 errors, ANOVA allows to simultaneously do this in one test. Overall, the One-Way ANOVA is a rebuttal to the independent samples t-test and provides a solution to overcome the dichotomous variable issue and extends itself to assessing 'average differences' of more than two groups, like a race variable. This statistical test allows us to examine 'average differences' of multiple groups that are independent of each other by showing us exactly where the average group differences lie. One-way ANOVA allows us to work with variables with multiple response attributes and non-dichotomous type variables and through post hoc testing tell us which groups have a sizeable significant average difference(s). While ANOVA is mathematically complicated, in SPSS it is fruitful because recoding is at a minimum in this test. The variables are best used in their natural or original state and there are no concerns about transforming variables into smaller response attributes. One of the shortfalls of this test is that it only allows for only one main effect, making it overly parsimonious, compared to other ANOVA family statistical tests, like two-way factorial ANOVA. Other ANOVA family tests should be considered to overcome the issue of simplicity. The next chapter discusses two-way factorial ANOVA.

Keywords and Definitions

ANOVA	Analysis of variances and this test compares averages across many groups (i.e., non-dichotomous).
One-way ANOVA	Also identified as a comparison of mean test hypothesis testing bivariate statistical analysis examining an IV and DV. Here, the IV is referred to as a factor or main effect.

Main effect	Examines the independent of effects of each factor as it relates to the DV or outcome, regardless of other parameters or factors.
Homogeneity of Variance Test or Levene's Test for Equality of Variances	Examines if variances are equal or unequal and fulfils a core assumption.
Null hypothesis	This states that there are no average differences amongst the groups.
Research hypothesis	This states that there are average differences amongst the groups.
F-statistic	This is the sample statistic that tells if there are significant average differences between the groups.
F-welch	This is the sample statistic correction when the variance assumption of equal variances is violated.
Eta^2	This is the magnitude of effect size and how much of variability in the DV is accounted for by a single main effect or factor or IV.
Post-hoc testing	This tells us where the exact significant average differences amongst groups exist. Tukey, Scheffé, Dunnett's, and Bonferroni are all types of post-hoc tests.

Test Your Knowledge

1   _____is an extension of the independent samples t-test which takes on more than two group means (i.e., non-dichotomous variables).

   a   Independent samples t-test
   b   One-way ANOVA
   c   Paired samples t-test
   d   Post-hoc testing
   e   Averages

2   The ____ is the independent effect of the IV or factor on the outcome variable

   a   Interaction effect
   b   Replication effect
   c   Main effect
   d   Minor effect
   e Eta^2

3   The major difference between independent samples t-test and One-way ANOVA is:

   a   One has post hoc testing
   b   One has a dichotomous variable, and the other does not
   c   One has an interval-ratio DV
   d   One has one IV
   e   a and b

4 The research hypothesis in a One-way ANOVA states the following:

   a   The means are equal and there are no average differences
   b   The means are different and there are average differences in at least two of
       the groups
   c   The amount of variability in the DV is accounted for by the main effect
   d   The exact average differences amongst groups
   e   None of the above

5 The goal of a post hoc test is to

   a   Reject the null hypothesis
   b   Fail to reject the null
   c   Make note of exact significant average differences amongst groups
   d   Discuss variability
   e   a and c

6 The ANOVA family is a big family and has many extended family members like:
   Two-way ANOVA, ANCOVA, MANOVA and MANCOVA.

   a   True
   b   False

7 The assumption of variances in One-way ANOVA should be ____ and if not the
   ____ should be utilized as a correction sample statistic.

   a   Equal
   b   Variance
   c   Unequal
   d F_{welch}
   e   F-statistic (original)

8 One-way ANOVA has taught us the following things:

   a   It compares of means of many groups
   b   It examines average differences
   c   It provides magnitude of effect size
   d   It maintains Equal Variances
   e   All of the above

9 ____ is the most robust post-hoc testing procedure.

   a   Tukey
   b   Scheffé
   c   Bonferroni
   d   F-statistic
   e   Levane's Test of Equality of Variances

10 One-way ANOVA assesses interaction effects.

   a   True
   b   False

8 Multivariate Hypothesis Testing Using Comparison of Means Using Two-Way Factorial ANOVA

Sometimes, in doing real world analysis, the scope of the research design for comparison of means extends beyond univariate ANOVA designs. Our primary interest becomes more than just single variable analysis or main effect for hypothesis testing and comparison of means on a single DV. Sometimes research designs are more complex and multilevel and require a different type of ANOVA test. Most often in the social sciences, there are multiple factors that contribute to the explanation of an outcome. Seldom there is one factor that is responsible for a particular outcome. For example, a researcher wants to examine two factors or main effects with a single DV, like race and political party affiliation (factors or main effects) and # of weapons possessed (the DV). Here, the interest is in examining how racial background and political party affiliation, liberal, conservative, or independent, impact average differences in # of weapons possessed. Both the main effects and interaction effect shed light on this question of # of weapons possessed. In the social world, we know that your racial background, along with your political party affiliation may effect # of weapons possessed and the question of interest is: Are there statistically significant 'average or mean differences' in # of weapons possessed based on racial background and political party affiliation?

Two-way factorial ANOVA statistical research design is between group (i.e., different people in each group) analysis of variance. Here, the two represents two factors or main effects of IVs to compare average difference between groups. The two-way ANOVA is a parametric test; it has several assumptions that must be met so the results are meaningful and powerful.

In this type of ANOVA analysis, the independent effects of each factor's contribution to the model are assessed, as well as the combined effects. Thus, a Two-way factorial ANOVA provides us with the following:

1   One main effect for variable A (A)
2   One main effect for variable B (B)
3   One interaction effect for variable A*variable B (A*B)

Recall, that the main effect is the independent effect a factor has on the DV. Then the question that peaks our curiosity is: What is an interaction effect?

An interaction effect is the combined effects of both factors on the outcome. Specifically, it means that the effect of one IV on the outcome variable depends on the second factor or IV. They are not independent of each other, but rather contingent upon one another. One-way ANOVA did not cover this, as there is no interaction, only a single main effect.

DOI: 10.4324/9781003215691-10

This type of statistical analysis introduces a second factor to the model and analyzes the independent effects of both factors, as well the interaction or combined effects. Here, the interaction effect always takes precedence over the main effects if it is statistically significant. The interaction effect becomes the focal point of discussion with regards to the DV.

In a Two-way factorial ANOVA, you are most often assessing the following scenarios:

- Statistical differences between the means of two or more groups in the main effects for factor A and factor B, as well as the interaction effect of factor A*B
- The other statistical information that a Two-way ANOVA provides is explanatory power of the model by assessing estimate size through eta^2 and adjusted R^2 values.

What variables do you need to run a two-way factorial ANOVA?

- At least two IVs or factors: Two nominal or ordinal categorical or discrete variables with more than two response attributes.
- Dependent variable: One interval-ratio continuous or scaled.

Assumptions	Explanation
1. Data or level of measurement of variables	The data or level of measurement assumption clearly outlines the types of variables a two-way factorial ANOVA is bound to. In this an IV that is nominal or ordinal categorical or discrete is key and a DV that is interval-ratio continuous or scaled
2. Sampling method and sample size	Random sampling and large sample
3. Independence of observations	Data observations must be independent of each other; no influence by other observations
4. Shape of the distribution	Normally distributed data
5. Homogeneity of variances	Equal or not equal variances assumed, where a preference is given to 'equal' variances. If the F-statistic is significant we report not equal variances; if the F-statistic is not significant we report unequal or not equal variances
6. Confidence level	The amount of risk you are willing to take: 95%, 0.05 alpha
7. Sample statistic	F-statistic or F-ratio: rejects the H_o or fails to reject it.
8. Outliers	Minimize outliers or extreme scores

In two-way factorial ANOVA the research question is written in specific terms and must include: the two main effects and the interaction effect. The research question and the null and research hypothesis are written in specific terms. Let's see the example below:

Research Question: There are no statistically significant average differences in the DV for the main and interaction effects?

Null and research hypotheses

Null Hypotheses:

- There are no statistically significant average or mean differences in the DV for the main effect, IVa/factor A
- There are no statistically significant average or mean differences in the DV for the main effect, IVb/factor B
- There is no statistically significant interaction between factor * factor B for the DV

Example: There are no statistically significant average differences between the main effect of race on number of times being convicted (factor A); There are no statistically significant average differences between the main effect of interest in school on number of times being convicted (factor B); there is not statistically significant interaction between race*interest in school on number of times being convicted.

Research or alternate hypothesis:

There are statistically significant average differences between the main effect of race on number of times being convicted (factor A); There are statistically significant average differences between the main effect of interest in school on number of times being convicted (factor B); there is statistically significant interaction between race*interest in school on number of times being convicted.

- There are statistically significant average or mean differences in the DV for the main effect, IVa/factor A
- There are statistically significant average or mean differences in the DV for the main effect, IVb/factor B
- There is statistically significant interaction between factor * factor B for the DV

These research hypotheses for two-way factorial ANOVA can be visualized like so,

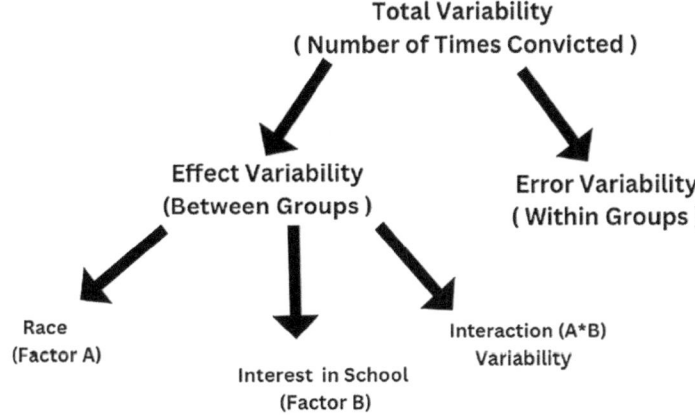

Figure 8.1 Tree Diagram of a Two-Way Factorial ANOVA of each Main Effect, Race and Interest in School and the Combined Effects or Interaction Effect

The sample statistic here is like One-way ANOVA, it is the F-statistic or F-ratio. This statistic tells us if we reject or fail or partially reject the null hypothesis. However, this time three F-statistics' are calculated for each effect. There is one for main effect A, B, and the interaction effect. Please note that if the interaction effect F-statistic is statistically significant that effect supersedes the main effect(s) and thus becomes the focal point of the analysis. It takes precedence. If all three effects are statistically significant you have full rejection of the model; if only two out of three or one out of three effects are statistically significant then there is only partial significance claimed and partial rejection of the null hypothesis.

Two-Way Factorial ANOVA and Post-hoc Testing

The F-statistic sample statistic only highlights an average or mean difference, but does not inform us of which groups significantly differ or not. To establish precision in mean differences, a post-hoc test needs to be run to establish exactly where the group differences lie. This is important to know. Knowing that there are mean differences is great, but to complete the analysis reporting of which groups significantly differ is equally important to this type of analysis. There are many post-hoc testing procedures based on the research design and nature of data, like Tukey, Scheffé, Dunnett's, and Bonferroni, that the ANOVA family offers. Bonferroni post-hoc test is the most robust and popular one that is generally used. Post-hoc testing is an integral part of the Two-way factorial ANOVA test and must be discussed for a complete analysis of results.

Effect Size and Two-Way ANOVA

In the context of a Two-way factorial ANOVA, effect sizes measures quantify the practical significance of observed average differences. Two-way ANOVA produces two calculations of effect size. The first is the partial eta^2. The effect size tells how much contribution each main effect and interaction effect to the DV, independent of each other. An eta^2 is calculated to explain what proportion of the variance in the DV is being explained by both main effects and the interaction effect or factors or IVs. The closer to a 100% explanation, the better the explanation. No research finding will give a 100% explanation, but the closer the value is to 100 the better the explained variance. These can be explained by weak, moderate, or strong effects. The Adjusted R^2 tells us how much variation in the DV is accounted for by the overall model, the main and interaction effects combined. An eta^2 is calculated to explain how much variance in the DV is being explained by the main effect or factor or IV. The closer to a 100% explanation, the better the explanation. Effect size speaks to the explanatory power of the model out of 100% and must be noted and discussed. Again, the closer the value is to 100 the better the model. Low percentages are indicative of weak factors, main effects, and interaction effects and therefore reconceptualization or conceptual re-specification of the model is needed or at most, needs to be discussed further. A large effect size indicates that there is greater practical significance in the average differences reported.

Key Differences Amongst One-Way ANOVA vs. Two-way Factorial ANOVA

	One-way ANOVA	*Two-way factorial ANOVA*
Variables	One IV with more than two groups that are nominal or ordinal	Two IVs with more than two groups that are nominal or ordinal
Effects	One main effect + no interaction effect	Two main effects + one interaction effect
Homogeneity of variance	Uses F_{welch} test if variances are not equal	Must discuss in limitations
Explanatory power	None in One-way ANOVA	Partial eta^2 and Adjusted R^2

Key Statistics to Report

1   Descriptive statistics: Report basic Descriptive statistics, sample size (N), averages, grand mean, standard deviation, minimum and maximum values; provide trends and patterns of data; this provides readers with a quick overview of each group characteristics as it relates to the DV or social phenomenon by each factor.
2   Levene's Test of Homogeneity of Variances: Report whether equal or not equal variances are assumed; Equal ($p > 0.05$) or not equal variances ($p < 0.05$): Tests

the assumption of equal or not equal variances, remember you can never have two nots together (i.e., not significant, means equal variances; significant means not equal variances); non-violation of the assumption yields equal variances amongst groups.

3   F-value or F-ratio or F-statistic: Report the calculated test or sample statistic that evaluates statistical significance to reject or fail to reject or partial rejection of the null hypothesis of overall comparison of means; This value report significant mean differences or not; degrees of freedom can be reported but is not mandatory.

4   Partial eta^2 (effect size) and Adjusted R^2: magnitude of effect size and reports the proportion of total variance attributed to group differences and whether a meaningful difference is apparent for each factor independent of each other and the overall full model using Adjusted R^2. R^2 can be reported, it is the less conservative number.

5   Post-hoc testing: Report specific significant average or mean group differences. Bonferroni is the most robust test. Other post-hoc test may also be run.

6   Profile plots of means which visually depict the averages for each main effect and interaction effect

7   Always tell the reader, what the confidence level and alpha is set to. For example, is it at 95%, alpha of 0.05 or 99%, alpha of 0.001?

Two-way factorial ANOVA indeed is the better comparison of means test as it provides the most information about 'average differences'. This comparison of means hypothesis testing technique overrides the independent samples t-test and One-way ANOVA as it has great explanatory power about average group differences across two main and interaction effects; moreover, it produces effect sizes for all main and interaction effects, as well as provides a value for the overall explained variance for the total model. One-way ANOVA and Two-way factorial ANOVA only assess one DV and don't assess control variables. Other tests of the ANOVA family do this, like, ANCOVA, MANOVA and MANCOVA. Two-way factorial ANOVA, while an interesting statistical tool, has limitations. First, it is heavily assumption dependent. Assumptions must be met to yield reliable and valid outcomes. This is important. Failure to meet assumptions results in faulty findings. Sometimes, the interaction term becomes complex to evaluate completely and requires more detailed observations and analysis for each main effect, especially when the interaction term deems statistically significant. Large sample sizes are needed for adequate and reasonable results. Small sample sizes may be misleading to the analysis. Each main effect should have equal number of observations for the levels of the factor. Unequal observations may lead to biased results. Lastly, extreme outliers may negatively influence the outcomes of two-way factorial ANOVA. Recognize outliers and normalize data patterns using data transformation techniques.

Activity Alert

Carefully, list all assumptions associated with the Two-way ANOVA.
Should variances be equal or not for the homogeneity of variance test?
What is the key difference(s) between One-way ANOVA or Two-way ANOVA?
What are the core statistics to report in this test? What are the unique statistics of this test that are not seen in one-way ANOVA or independent samples t-test?

Source of Variation	d.f.	SS	MS	F_0
Factor A (between groups)	a-1	$SSA = \sum_{i=1}^{a} n_i \left(\bar{y}_{i.} - \bar{y}_{..} \right)^2$	$MSA = \dfrac{SSA}{(a-1)}$	$\dfrac{MSA}{MSE}$
Factor B (between groups)	b-1	$SSB = \sum_{j=1}^{b} n_j \left(\bar{y}_{.j} - \bar{y}_{..} \right)^2$	$MSB = \dfrac{SSB}{(b-1)}$	$\dfrac{MSB}{MSE}$
Error (within groups)	(a-1)(b-1)	$SSE = SST - SSA - SSB$	$MSE = \dfrac{SSE}{(a-1)(b-1)}$	
Total	N-1	$SST = \sum_{i=1}^{a} \sum_{j=1}^{n} \left(y_{ij} - \bar{y}_{..} \right)^2$		

Figure 8.2 Two-Way Factorial ANOVA Formula

Two-way Factorial ANOVA in SPSS

1 Click on Analyze > General Linear Model > Univariate (Two-Way Factorial ANOVA)
2 Move the variables of interest into the Dependent List (DV) and Fixed Factors List (IVs)
3 Click on Post Hoc >Check off Bonferroni
4 Click on Options >Check off Descriptives, Homogeneity of Variance Test, and Estimates of Effect Size
5 Click on EM Means->Place Overall Means into Display Means for 'Overall' into Box
6 Click on Profile Plots to Generate Plots (the factor with less response attributes goes into separate lines and the other goes to horizontal axis) and then ADD
7 OK to generate a Two-Way Factorial ANOVA

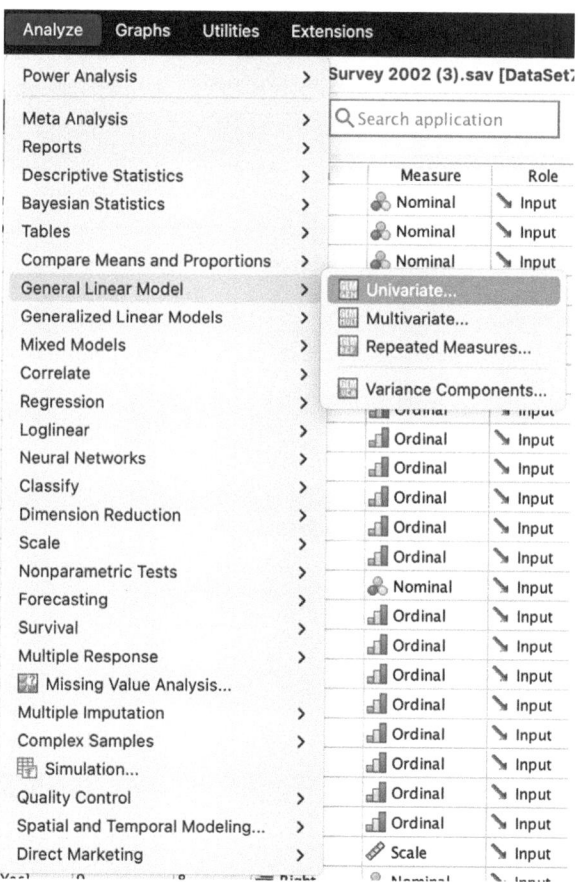

Figure 8.3a SPSS Command for Two-Way Factorial ANOVA

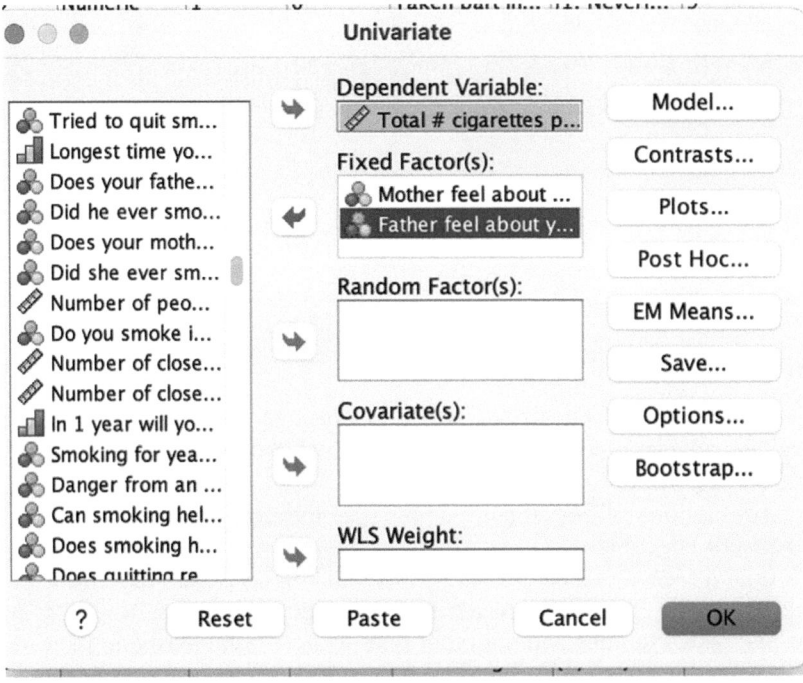

Figure 8.3b Add the DV to the Dependent Variable Box and the two IVs to the Fixed Factor Box

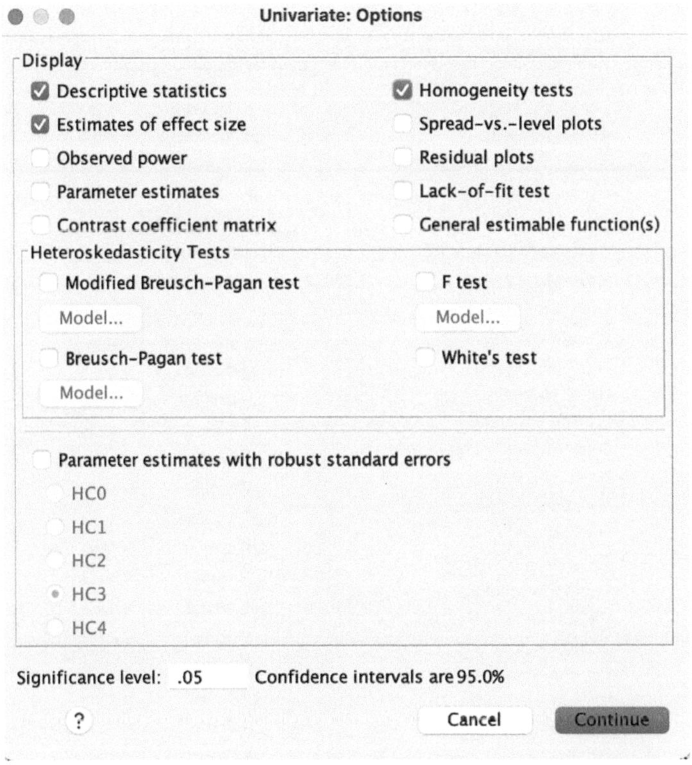

Univariate: Options

Display

☑ Descriptive statistics   ☑ Homogeneity tests

☑ Estimates of effect size   ☐ Spread-vs.-level plots

☐ Observed power   ☐ Residual plots

☐ Parameter estimates   ☐ Lack-of-fit test

☐ Contrast coefficient matrix   ☐ General estimable function(s)

Heteroskedasticity Tests

☐ Modified Breusch–Pagan test   ☐ F test

  Model...   Model...

☐ Breusch–Pagan test   ☐ White's test

  Model...

☐ Parameter estimates with robust standard errors

  ○ HC0

  ○ HC1

  ○ HC2

  ⦿ HC3

  ○ HC4

Significance level: .05   Confidence intervals are 95.0%

?   Cancel   Continue

Figure 8.3c In Options Check-off Descriptive Statistics, Estimates of Effect Size and Homogeneity of Tests

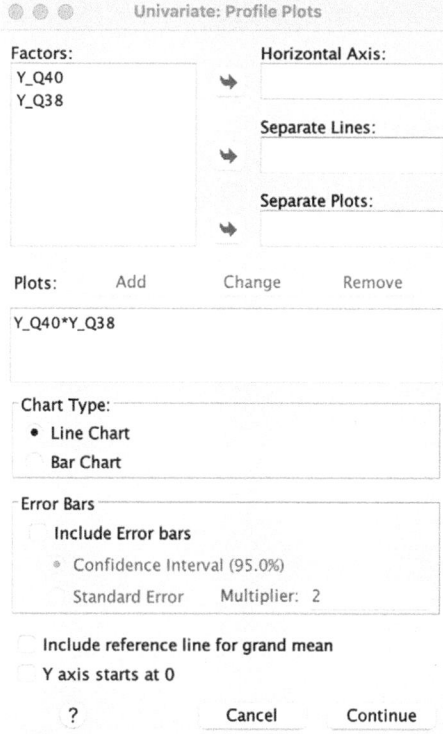

Figure 8.3d Add a Variable to the Horizontal Axis and the Second one to Separate Lines and Click Add

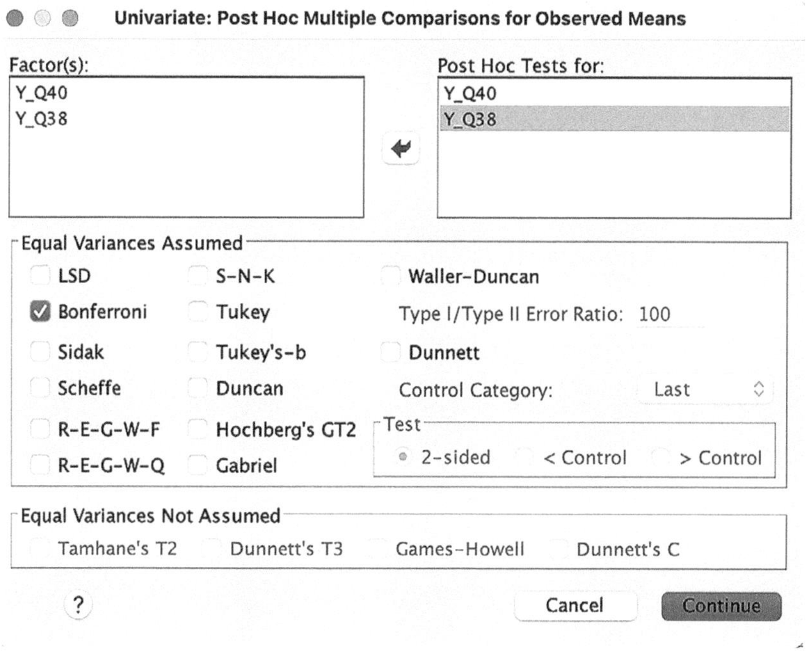

Univariate: Post Hoc Multiple Comparisons for Observed Means

Factor(s):

Y_Q40
Y_Q38

Post Hoc Tests for:

Y_Q40
Y_Q38

Equal Variances Assumed

- LSD
- ☑ Bonferroni
- Sidak
- Scheffe
- R-E-G-W-F
- R-E-G-W-Q

- S-N-K
- Tukey
- Tukey's-b
- Duncan
- Hochberg's GT2
- Gabriel

- Waller-Duncan

Type I/Type II Error Ratio: 100

- Dunnett

Control Category: Last

Test
- ◦ 2-sided   < Control   > Control

Equal Variances Not Assumed

- Tamhane's T2   Dunnett's T3   Games-Howell   Dunnett's C

? | Cancel | Continue

Figure 8.3e Add Both Variables to the Post-Hoc Tests Box to the Right and Check-off Bonferroni

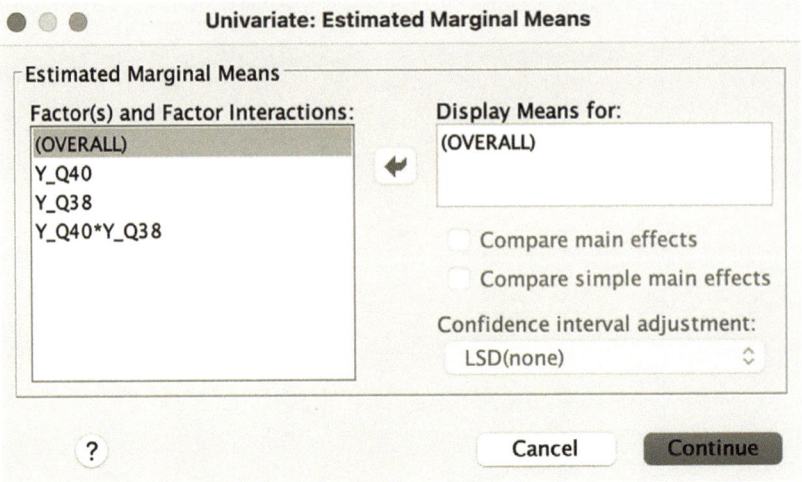

Figure 8.3f Display Means for Overall Model

Between-Subjects Factors

		Value Label	N
Mother feel about your smoking	1	She approves	25
	2	She doesn't care	104
	3	She doesn't like it	255
	4	She doesn't know that I smoke	333
Father feel about your smoking	1	He approves	15
	2	He doesn't care	111
	3	He doesn't like it	209
	4	He doesn't know that I smoke	382

Figure SPSS Output #8 Comparison of Means Two-Way Factorial ANOVA

Descriptive Statistics

Dependent Variable: Total # cigarettes past 7 days

Mother feel about your smoking	Father feel about your smoking	Mean	Std. Deviation	N
She approves	He approves	64.91	55.199	11
	He doesn't care	83.33	39.577	3
	He doesn't like it	59.00	37.568	7
	He doesn't know that I smoke	51.25	44.642	4
	Total	63.28	45.639	25
She doesn't care	He approves	48.00	.	1
	He doesn't care	68.42	43.535	64
	He doesn't like it	54.48	40.091	27
	He doesn't know that I smoke	39.67	22.158	12
	Total	61.29	41.441	104
She doesn't like it	He approves	121.67	78.258	3
	He doesn't care	49.83	37.587	36
	He doesn't like it	39.99	34.192	168
	He doesn't know that I smoke	28.00	29.493	48
	Total	40.08	35.947	255
She doesn't know that I smoke	He doesn't care	25.75	18.599	8
	He doesn't like it	35.29	28.064	7
	He doesn't know that I smoke	14.18	24.439	318
	Total	14.90	24.574	333
Total	He approves	75.13	60.415	15
	He doesn't care	59.72	41.961	111
	He doesn't like it	42.34	35.180	209
	He doesn't know that I smoke	17.10	26.162	382
	Total	32.27	37.057	717

Levene's Test of Equality of Error Variances[a,b]

		Levene Statistic	df1	df2	Sig.
Total # cigarettes past 7 days	Based on Mean	8.802	13	702	<.001
	Based on Median	7.355	13	702	<.001
	Based on Median and with adjusted df	7.355	13	603.730	<.001
	Based on trimmed mean	9.004	13	702	<.001

Tests the null hypothesis that the error variance of the dependent variable is equal across groups.

  a. Dependent variable: Total # cigarettes past 7 days

  b. Design: Intercept + Y_Q40 + Y_Q38 + Y_Q40 * Y_Q38

Tests of Between-Subjects Effects

Dependent Variable: Total # cigarettes past 7 days

Source	Type III Sum of Squares	df	Mean Square	F	Sig.	Partial Eta Squared
Corrected Model	274293.270[a]	14	19592.376	19.401	<.001	.279
Intercept	200182.472	1	200182.472	198.224	<.001	.220
Y_Q40	19680.366	3	6560.122	6.496	<.001	.027
Y_Q38	14603.985	3	4867.995	4.820	.002	.020
Y_Q40 * Y_Q38	16405.829	8	2050.729	2.031	.041	.023
Error	708934.697	702	1009.878			
Total	1729969.000	717				
Corrected Total	983227.967	716				

a. R Squared = .279 (Adjusted R Squared = .265)

Post Hoc Tests

Mother feel about your smoking

Multiple Comparisons

Dependent Variable: Total # cigarettes past 7 days
Bonferroni

(I) Mother feel about your smoking	(J) Mother feel about your smoking	Mean Difference (I–J)	Std. Error	Sig.	95% Confidence Interval Lower Bound	Upper Bound
She approves	She doesn't care	1.99	7.079	1.000	-16.74	20.72
	She doesn't like it	23.20*	6.660	.003	5.58	40.82
	She doesn't know that I smoke	48.38*	6.590	<.001	30.94	65.81
She doesn't care	She approves	-1.99	7.079	1.000	-20.72	16.74
	She doesn't like it	21.21*	3.697	<.001	11.42	30.99
	She doesn't know that I smoke	46.39*	3.570	<.001	36.94	55.83
She doesn't like it	She approves	-23.20*	6.660	.003	-40.82	-5.58
	She doesn't care	-21.21*	3.697	<.001	-30.99	-11.42
	She doesn't know that I smoke	25.18*	2.644	<.001	18.18	32.18
She doesn't know that I smoke	She approves	-48.38*	6.590	<.001	-65.81	-30.94
	She doesn't care	-46.39*	3.570	<.001	-55.83	-36.94
	She doesn't like it	-25.18*	2.644	<.001	-32.18	-18.18

Based on observed means.
The error term is Mean Square(Error) = 1009.878.

*. The mean difference is significant at the .05 level.

Father feel about your smoking

Multiple Comparisons

Dependent Variable: Total # cigarettes past 7 days
Bonferroni

(I) Father feel about your smoking	(J) Father feel about your smoking	Mean Difference (I–J)	Std. Error	Sig.	95% Confidence Interval Lower Bound	Upper Bound
He approves	He doesn't care	15.41	8.742	.470	-7.72	38.54
	He doesn't like it	32.79*	8.495	<.001	10.32	55.27
	He doesn't know that I smoke	58.03*	8.365	<.001	35.90	80.16
He doesn't care	He approves	-15.41	8.742	.470	-38.54	7.72
	He doesn't like it	17.38*	3.732	<.001	7.51	27.26
	He doesn't know that I smoke	42.62*	3.427	<.001	33.55	51.68
He doesn't like it	He approves	-32.79*	8.495	<.001	-55.27	-10.32
	He doesn't care	-17.38*	3.732	<.001	-27.26	-7.51
	He doesn't know that I smoke	25.24*	2.734	<.001	18.00	32.47
He doesn't know that I smoke	He approves	-58.03*	8.365	<.001	-80.16	-35.90
	He doesn't care	-42.62*	3.427	<.001	-51.68	-33.55
	He doesn't like it	-25.24*	2.734	<.001	-32.47	-18.00

Based on observed means.
The error term is Mean Square(Error) = 1009.878.

*. The mean difference is significant at the .05 level.

Profile Plots

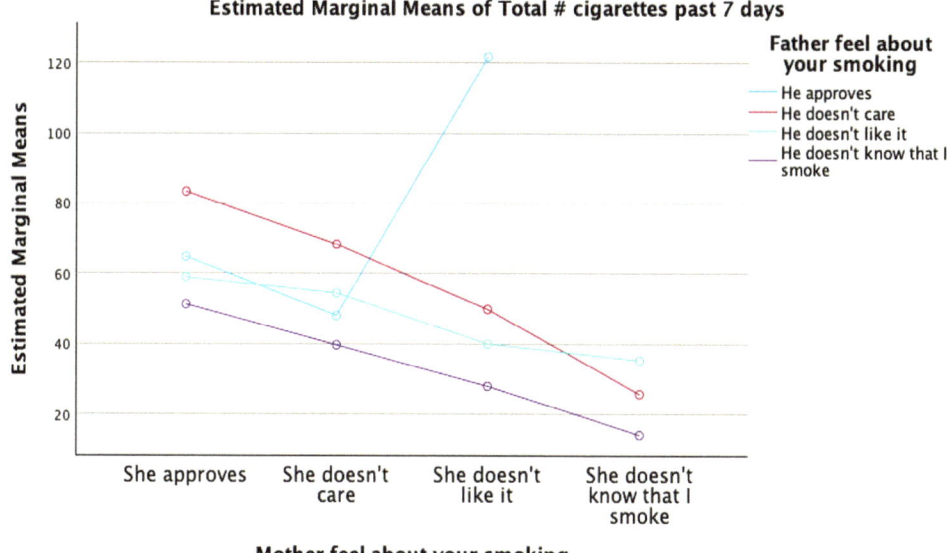

Estimated Marginal Means of Total # cigarettes past 7 days

Mother feel about your smoking

SPSS Example: Technical and Substantive Interpretation

Comparison of Means Hypothesis Testing: Two-Way Factorial ANOVA

Research Question: Are there average or mean differences in the main effects of how mother and father feel about your smoking on total number of cigarettes smoked in past seven days?

Null Hypothesis: There are no statistically significant average or mean differences seen in the main effects of how mother (A) and father (B) feel about your smoking and the interaction effect (A*B) on total number of cigarettes smoked in past seven days? The means are equal.

Research Hypothesis: There are statistically significant average or mean differences seen in the main effects of how mother (A) and father (B) feel about your smoking and the interaction effect (A*B) on total number of cigarettes smoked in past seven days? The means are different.

Technical interpretation

Descriptive statistics: The sample size for factor A and B, how mother and father feel about your smoking depict the following sample outcomes for mothers and fathers both (N = 715): 25 vs 15 approve; 104 vs 111 do not care; 255 vs 209 don't like it; and 333 vs 382 don't know that their kid smokes. Most parents did not know their child smokes and a handful approve smoking. The grand means for each group approve, doesn't care, doesn't like it, and doesn't know that I smoke are 63.28, 61.29, 40.08 and 32.27 number of cigarettes smoked in past seven days. The trends and patterns indicate that on average the greatest number of cigarettes smoked, about 122, in the past seven days is for those respondents in which mother did not like it, but there is paternal approval and the least number of cigarettes, about 14, on average, is when both parents do not know that the child smokes. Most groups scored above or below the grand mean, but number of cigarette smoking events in the past seven days is most often higher for respondents in which at least one parent approves.

Levene's Test for Equality of Variances: Levene's test for Equality of Variances suggests a statistically significant outcome (F = 8.802, $p < 0.05$) and therefore, equal variances are not assumed. Because equal variances results must be dealt with caution when making inferences.

Test statistic, Effect size and Post-hoc test: The main effect (A), how mother feels about your smoking, indicates an F-statistic that is statistically significant (F = 6.496, $p < 0.05$) suggestive of significant average group differences. The means are significantly different. Similarly, the second main effect (B), how father feels about your smoking also results in a statistically significant outcome (F = 4.820, $p < 0.05$) suggesting that there are average group differences. Additionally, the interaction effect of the combined effects (A*B) is also statistically significant indicating that in assessing 'average' group differences, both factors A and B must be explored simultaneously and in combination. The interaction effect now becomes the focal point of analysis and supersedes the previous main effects (A and B). There is a full rejection of the null hypothesis and thus there are significant mean or average differences amongst the groups for both factors or IVs. The effect size, partial eta squared indicates that the main effects of how mother and father feel explain about 3% and 2% of the variance in the number of cigarettes smoked in the past seven days. The interaction effect explains about 2.3%. The overall model or the adjusted R^2 explains about 27% of the variance in the model. The profile plot shows this interaction effect. The post-hoc test, Bonferroni, reveals exactly where the mean differences lie.

For mothers, the following groups demonstrated significant average differences ($p < 0.05$):

- She approves – she doesn't like it; she doesn't know that I smoke
- She doesn't care – she doesn't like it; she doesn't know that I smoke
- She doesn't like it – she doesn't know that I smoke

There was only one group that was not statistically significant: she approves and she doesn't care (1.99, $p > 0.05$).

For fathers, all combinations, except one group, demonstrated significant average differences ($p < 0.05$):

- He approves – he doesn't like it; he doesn't know that I smoke
- He doesn't care – he doesn't like it; he doesn't know that I smoke
- He doesn't like it – he doesn't know that I smoke

There was only one group that did not show average significant differences: he approves and he doesn't care (15.41, $p > 0.05$).

Substantive interpretation

In summary, Two-way factorial ANOVA demonstrated that how a mother, combined with father's feeling about smoking resulted in varying mean differences in the number of cigarettes smoked in the past seven days. When approval by both parents is there or not, average smoking in past seven days varies. The largest mean differences amongst mothers and fathers were when she approves and doesn't know that the child smokes and when fathers approve and do not know that child smokes. This Two-way ANOVA clearly shows that parental feelings about their child smoking matter, not for one but both parents. Their varying approval or disapproval impacts, on average, the total number of cigarettes smokes in past seven days. In the social world, parental approval matters, substance use is not unique and varies on how a parent views this. These results should be taken with caution as some groups had larger representation and sample sizes, also equal variances were not met, so inferences should be made cautiously.

Final Thoughts

The comparison of the means test, Two-way factorial ANOVA, is quite unique in what it has to offer. Even though we are still assessing 'average' or 'mean' differences, how they are being analyzed differs greatly, especially based on the information it generates. The Two-way factorial ANOVA is one of the more complicated ANOVA family tests. It has great explanatory power because it covers two main effects and one interaction effect explaining the outcome or DV. The independent samples t-test or One-way ANOVA did not do this – it was quite basic, comparatively speaking. The Two-way factorial ANOVA comes out as a strong contender for 'average differences' in the bigger picture of data analysis. It is at an advantage, compared to the other comparison of mean tests because it allows for greater explanations regarding the DV. To allow for the examination of interactions amongst factor or IVs is statistically valuable. This aids in the understanding of relationships assessing average differences and the combined effects of each factor. Main effects alone cannot provide such valuable insights. The two-way factorial ANOVA design allows to study complex multiple factor or IVs and in a way controls for potential confounding variables. The reliability and validity of such a test is enhanced tremendously. Instead of running multiple One-way ANOVAs, this method means researchers can explore both variables simultaneously. This is time efficient. Compared to the previous tests discussed, Two-way factorial ANOVA undoubtedly provides a concise and comprehensive outlook regarding the relationships being tested. Consideration of both main and interaction effects beautifies the discussion of 'average differences' or 'mean differences' which leads to an in-depth and holistic understanding of the social phenomenon at play. The statistical power that is seen in Two-way factorial ANOVA cannot be compared to One-way ANOVAs, if no assumption is violated. By providing key model fit tests, like partial eta^2 and Adjusted R^2, two-way factorial ANOVA outdoes itself in the world of comparison of means testing.

Keywords and Definitions

Two-way factorial ANOVA	A comparison of means test that assesses two main effects and one interaction effect.
Main effect	The independent effect the factor or IV has on the DV.
Interaction effect	The combined effects both factors or IVs have on the DV.
Partial eta^2	This measures variance in the DV by each main effect and interaction effect out of 100%.
Adjusted R^2	This measures variance in the DV for the overall model out of 100%.

Test Your Knowledge

1   The _____ is responsible for determining the explained variance for each effect in a Two-way factorial ANOVA.

    a   Cox and Snell R^2
    b   Adjusted R^2
    c   Partial eta^2
    d   Interaction effects
    e   b and c

2   The following is true about the two-way factorial ANOVA:

    a   It's a parametric test
    b   It assesses two main effects only
    c   It assesses two main effects, and 1 interaction effect
    d   It discusses effect sizes, eta^2 and Adjusted R^2
    e   Only a and b

3   A two-way factorial ANOVA is most different from other comparison of means test, like independent samples t-test and One-way ANOVA because:

    a   It's a non-parametric test
    b   It has a null and research hypothesis
    c   It has an 'interaction' effect
    d   It compares means
    e   It is bivariate

4   A two-way ANOVA has many assumptions, however ____ assumption states that equal variances must be present.

    a   Normality
    b   Random sampling
    c   Homogeneity of variances
    d   Confidence level
    e   Sample statistic

5   The ____ is the sample statistic for the Two-way factorial ANOVA.

    a   t-value
    b   F-statistic or F-ratio
    c   Chi-Square
    d   Levene's F-statistic
    e   Cronbach's alpha

6   If assumptions are not me and violated what does this mean?

    a   It means that all generalizations must proceed with caution and inferences from a sample to a population need to be carefully taken and considered
    b   It means that generalizations can be confidently made
    c   It means that the null hypothesis is rejected
    d   a and b
    e   b and c

7 The ____ informs of the overall explained variance of the model.

    a   The adjusted R^2
    b   Partial eta^2
    c   t-value
    d   normality
    e   Homogeneity of variances

8 ____type of sampling in an assumption of the two-way factorial ANOVA

    a   Random
    b   Convenience
    c   Systematic
    d   Simple random sampling
    e   Purposive sampling

9 Group means are compared to the _____.

    a   Grand mean
    b   Large mean
    c   Average differences
    d   Statistically significant differences
    e   Partial eta^2

10 The central limit theorem guides the Two-way factorial ANOVA.

    a   True
    b   False

9 Multivariate Hypothesis Testing Using Comparison of Means by means of Advanced ANOVA Statistical Hypothesis Testing using ANCOVA, MANOVA, MANCOVA

The ANOVA family of tests discussed in these next sections are considered 'advanced' multivariate statistical techniques or analyses. They are not simple statistical techniques but rather are complex methods to engage in hypothesis testing techniques that compare means across multiple groups. These tests increase the all-inclusive explanations of relationships and their corresponding outcomes with and without covariates or control variables. Most often, their use is seen in advanced research designs or graduate courses. These tests are seldom discussed at the undergraduate level, however their overall importance in data analysis should not be underestimated, as they are powerful tests. Each technique is unique and serves a specific research agenda, provided the data complies. The more complex comparison of means test are: ANCOVA, MANOVA, and MANCOVA. Each test is presented and discussed in this order. Here are some similarities and differences about these advanced statistical tests:

ANCOVA	MANOVA	MANCOVA
Multivariate hypothesis testing	Multivariate hypothesis testing	Multivariate hypothesis testing
Significant average or mean differences while controlling for effects of covariate(s)	Significant average or mean differences on a set of DVs holistically with no covariate(s)	Significant average or mean differences on a set of DVs holistically, while also controlling for effects of covariate(s)
One or more categorical or discrete nominal or ordinal IV or factor with one interval-ratio continuous DV + covariate that is interval-ratio continuous	One or more categorical or discrete nominal or ordinal IV or factor with at least two interval-ratio continuous DV + no covariate	One or more categorical or discrete nominal or ordinal IV or factor with at least two interval-ratio continuous DV + covariate that is interval-ratio continuous

ANCOVA (Analysis of Covariance)

First, Analysis of Covariance or ANCOVA is an advanced multivariate ANOVA technique with more than two groups' averages being compared on a particular DV. It is a statistical analysis that combines components of ANOVA and regression. It basically evaluates whether population means of a DV are equal or the same across different levels of a categorical IV, while controlling for the effects of

DOI: 10.4324/9781003215691-11

continuous interval-ratio variable-covariates. ANCOVA lets researchers control for effects of one or more control variables or covariates. The diversification of ANCOVA is that it adds a control variable or covariate to the model and reduces the effects of any confounding variables on the analysis. For example, you can ask the following question and utilize ANCOVA to answer the question: Does level of education and racial background have, on average, varying # of arrests, after controlling for family support or geographical location (i.e., urban, rural). ANCOVA is an extension of the ANOVA analysis, with a bit of a twist. Here the main effects and interactions are assessed after the effects of some other variable, covariate, or concomitant variable, have been eliminated or removed or controlled for. The means across groups are *not* compared directly, but rather the means are *adjusted* by the covariate. Any variables that should theoretically correlate with the DV or that have been shown to correlate with the DV should be considered as the 'covariate'. ANCOVA is an improvement over the corresponding ANOVA model because it explains additional variability and thus results in greater precision, compared to a One- or Two-way factorial ANOVA model. The advantage of adding a covariate to the model increases precision of the calculated average or mean differences; moreover, it enhances statistical power of the model and makes it a more efficient analysis. So, what is a covariate? A covariate is a continuous interval-ratio variable or its equivalent (i.e., scaled) that controls for the effects on the DV. Its role is to minimize the effects of variability in the DV due to the covariate. The covariate and the DV should assume a linear and consistent relationship.

The key functionality of ANCOVA is it enhances sensitivity of the F-tests main effects and interaction effects by reducing error variance or 'systemic bias'; it improves research design when random assignment is not feasible; it controls for certain effects or variables that may impede that relationship; it works best when samples sizes between groups being tested are equal; ANCOVA runs the model and partial out the effects of the covariate by controlling for the main effects and interaction effects in the model (Mertler & Reinhart, 2017). Some real-life examples of ANCOVA are the following:

> Which advertising is most effective? An ice cream shop owner wants to know if advertising certain flavors, vanilla, strawberry, or mango effect # of customers who enter the shop. However, he is also aware that outside temperature may affect the # of customers. Therefore, his study compares customer traffic for the three different flavors with daily high temperature as a covariate.
>
> If you wanted to determine whether job satisfaction, gender, or the interaction between them influences income level (DV) you may want to eliminate or control for any other variable that may be attributable to income. Thus, a variable like education can be used as a covariate, which can be controlled for thus eliminating any effect, it may have on the DV, income.

Adding a covariate to any of these models will accurately determine if the average differences truly exist because of the factors or main effects and are not the influence of the covariate.

Main Differences in ANOVA vs. ANCOVA

Main difference	ANOVA	ANCOVA
Inclusion of covariate	No covariate or control	Adds covariate or control
Objective	Reports on significant average differences on a DV, with no covariate	Reports on significant average differences on a DV, while controlling for covariates; adjust means accordingly
Handling differences	Groups have similar baseline differences	Covariates play a pivotal role in adjusting for any baseline differences
Statistical power	Lower statistical power	Increased statistical power
Homogeneity of regression slopes	Does not take this into account	The relationship between the DV and covariate is similar across all levels of the IV(s) or factor(s)

What variables do you need to run an ANCOVA?

- At least one IV or factor: One nominal or ordinal categorical or discrete variable with more than two response attributes; you can always add a second factor or IV.
- Dependent variable: One interval-ratio continuous or scaled.
- Covariate: Ideally interval-ratio or its equivalent; but can be nominal or ordinal.

Remember, the results of ANCOVA are reliable and valid based on the assumptions that are clearly met and not violated.

Assumptions	Explanation
1. Data or level of measurement of variables	The data or level of measurement assumption clearly outlines the types of variables an ANCOVA is bound to a DV and covariate that is interval-ratio or its equivalent. In this an IV that is nominal or ordinal categorical or discrete with more than two groups
2. Sampling method and sample size	Random sampling and large sample
3. Independence of OBSERVATIONS	Data observations must be independent of each other; no influence by other observations. No relationship between or among groups
4. Shape of the DISTRIBUTION	Normally distributed data or normality of residuals or error terms for each category of the factor or IV. Shapiro-Wilk Test is used to assess this
5. Homogeneity of variances or homoscedasticity	Levene's Test for Equality of Variances: Equal or not equal variances assumed, where a preference is given to 'equal' variances that variances are equal across all levels of the IV, after accounting for effects of covariates. If the F-statistic is significant we report not equal variances; if the F-statistic is not significant we report unequal or not equal variances
6. Linearity	The covariate must be linearly related to the outcome or DV. Scatterplots can test this

Assumptions	Explanation
7. Slopes of regression lines	The slopes of the regression lines are parallel or homogeneous (slopes from same group); there should be no interaction between the IV or factor and the covariate; fail to reject the null hypothesis and thus should not be statistically significant. Build custom model, then run full model
8. Confidence level	The amount of risk you are willing to take: 95%, 0.05 alpha
9. Sample statistic	F-ratio or F-statistic rejects the H_o or fails to reject it
10. Outliers	Minimize outliers or extreme scores

How do you select a covariate for an ANCOVA? The answer lies in the conceptualization of the model or relationship. Most often any variable that conceptually and theoretically aligns or correlates with the DV is the selected covariate. The selected 'covariate' should be reliable and significantly correlated with the DV. A covariate or concomitant variable should ideally be measured at the interval-ratio continuous level of measurement. Demographic variables, at varying levels of measurement make ideal covariates for ANCOVA, but interval-ratio continuous variables are best. For example:

- Age
- Pretest scores
- IQ
- Race
- Gender
- Education

Another question that arises is how many covariates are too many? Basically, if the analysis is exhaustive and conclusive then, based on the research objective, covariates are tested for. No more than 1–3 covariates should be used, depending on the sample size.

In ANCOVA, the research question must include the factors or IV(s), the DV and the covariate variable. The research question and the null and research hypothesis are written in specific terms. Let's see the example below:

Research Question: Are there significant average differences in GPA in type of learning methods (traditional, hybrid, online), while controlling for number of courses taken.

Null hypothesis: There are no statistically significant average or mean differences in the DV for the groups of Factor A (main effects), while controlling for the concomitant/covariate.

There are no statistically significant average or mean differences in GPA for type of learning, while controlling for number of courses taken.

Research hypothesis: Are there significant mean differences for the DV between the groups of Factor A, after controlling for the concomitant variable?

There are statistically significant average or mean differences in GPA for type of learning, while controlling for number of courses taken.

The interpretation of ANCOVA is like other ANOVA family tests with the addition of a covariate to the model.

ANCOVA and Post-hoc Testing

The calculated sample statistic, F-statistic, only highlights an average or mean difference, but does not inform us of which groups significantly differ or not. To establish precision in mean differences, a post-hoc test needs to be run to establish exactly where the group differences lie. This is important to know. Knowing that there are mean differences is great, but to complete the analysis reporting of which groups significantly differ is equally important to this type of analysis. There are many post-hoc testing procedures based on the research design and nature of data, like Tukey, Scheffé, Dunnett's, and Bonferroni, that the ANOVA family offers. The Bonferroni post-hoc test is the most robust one and popular one that is generally used. Post-hoc testing is an integral part of the ANCOVA test and must be discussed for a complete analysis of results.

Effect Size and ANCOVA

The magnitude of effect size in ANCOVA to measure the true significance of observed differences is eta^2 for each independent factor(s) and the overall model. Again, like the other ANOVA family tests, eta^2 represents the variance explained in the DV by the factor(s) or main effect(s) and covariate(s) together. The closer this number is to 1, the larger the effect size. The closer to 0, the smaller the effect size. The Adjusted R^2 also provides overall effect of the entire model out of 100%. Again, this is an important measure of the strength of an effect and tells the proportion of explanation accounted in the DV for each factor(s) after accounting for the effects of the covariate(s).

Activity Alert

How does ANCOVA differ from One- and Two-way ANOVA?
What is a covariate?

Key Statistics to Report

1   Descriptive statistics: Report basic descriptive statistics, sample size (N), averages, grand mean, standard deviation, minimum and maximum values; provide trends and patterns of data; this provides readers with a quick overview of each group characteristics as it relates to the DV or social phenomenon by each factor.
2   Levene's Test of Equality of Variances: Report whether equal or not equal variances are assumed; Equal ($p > 0.05$) or not equal variances ($p < 0.05$). Tests the assumption of equal or not equal variances, remember you can never have two nots together (i.e., not significant, means equal variances; significant means not equal variances); non-violation of the assumption yields equal variances amongst groups. Equal variances are ideal.
3   The slopes of the regression lines: These lines are parallel or homogeneous (slopes from same group); there should be no interaction between the IV or factor

and the covariate; fail to reject the null hypothesis and thus should not be statistically significant.

4   F-value or F-ratio or F-statistic: Report the calculated test or sample statistic that evaluates statistical significance to reject or fail to reject or partial rejection of the null hypothesis of overall comparison of means; This value report significant mean differences or not; Degrees of freedom can be reported but is not mandatory.

5   Partial eta^2 (effect size) and Adjusted R^2: magnitude of effect size and reports the proportion of total variance attributed to group differences and whether a meaningful difference is apparent for each factor independent of each other and the overall full model using Adjusted R^2. R^2 can be reported, it is the less conservative number.

6   Post-hoc testing: Report specific significant average or mean group differences. Bonferroni is the most robust test. Other post-hoc test may also be run.

7   Profile: Plots of means which visually depict the averages for each main effect and interaction effect

8   Always tell the reader, what the confidence level and alpha is set to. For example, is it at 95%, alpha of 0.05 or 99%, alpha of 0.001?

ANCOVA in SPSS

1   Click on Analyze > General Linear Model > Univariate
2   Select Model and Build Terms to create main effects and interaction effects; interaction effect must be not significant to fulfill regression of slopes assumptions.
3   Move the variables of interest into the Dependent List (DV) and Fixed Factors List (IVs) and proceed to the Covariate Box and add the covariate of interest into the box
4   Click on EM Means >Transfer variable from the Factor(s) and Factor Interactions box to the Display means for Factor and check off compare main effects and select Bonferroni
5   Click on Univariate Options >Check off Descriptives, Homogeneity of Variance Test, and Estimates of Effect Size
6   OK to generate an ANCOVA

Shapiro Wilk in SPSS

1   Go to Analyze → Descriptive Statistics → Explore.
2   Move the variables you want to test for normality from the left-hand box to the right hand 'Dependent List' box….
3   Click Options….
4   Check Normality Plots with Tests.
5   click OK.

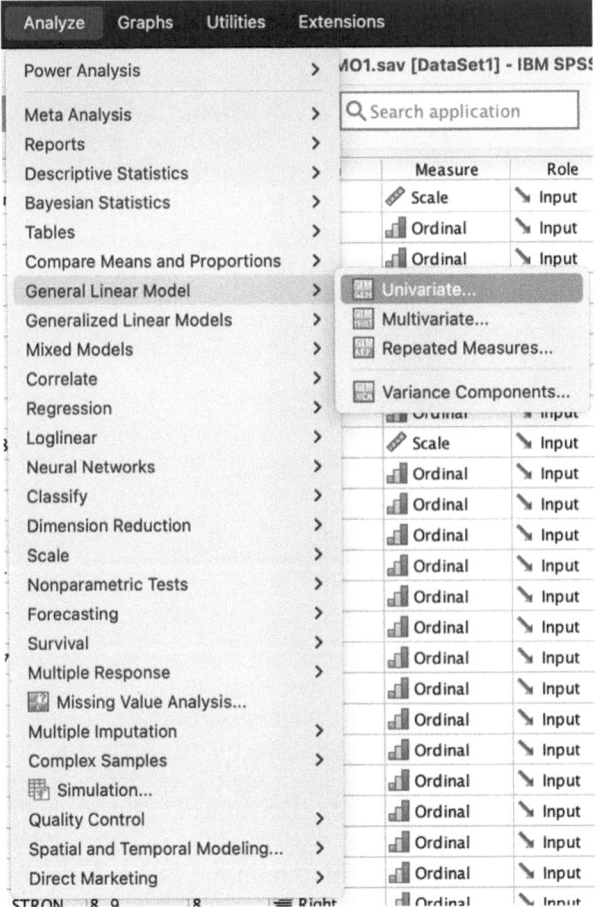

Figure 9.1a SPSS Command for ANCOVA

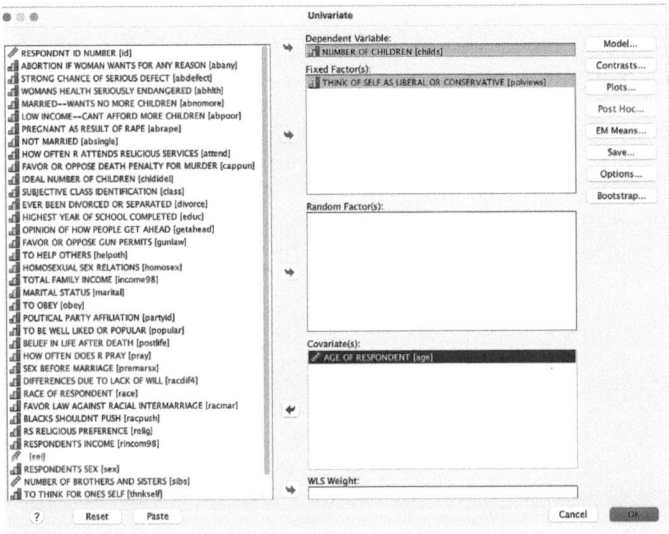

Figure 9.1b Add the DV to the Dependent Variable Box, the IV to the Fixed Factor Box, and the Covariate or Control Variable to the Covariate Box

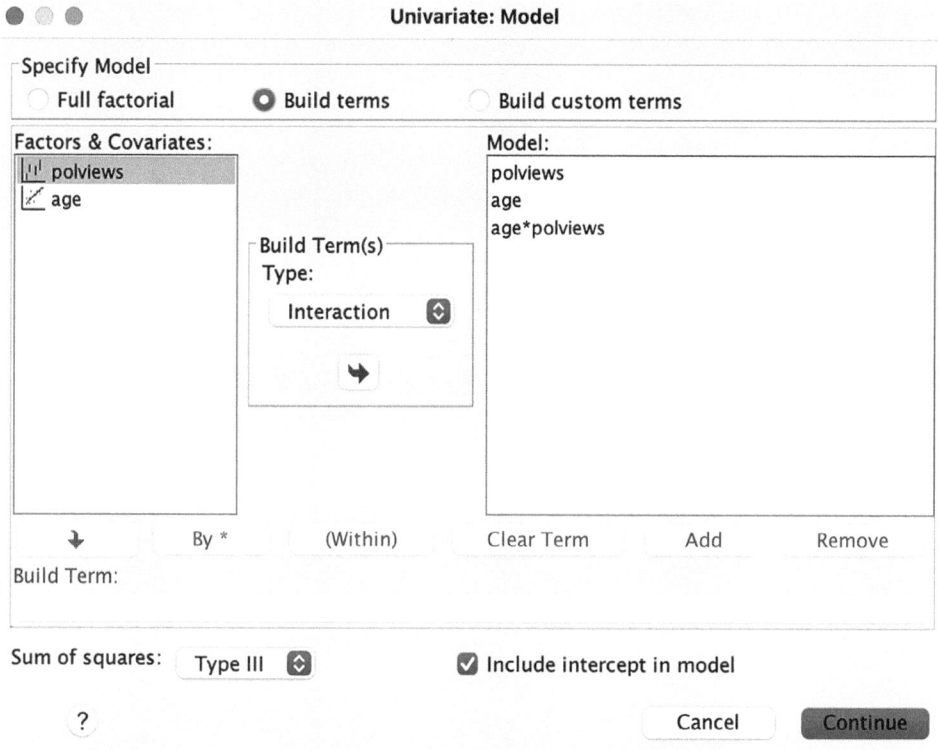

Figure 9.1c Specify Model and Build Terms to Create Main and Interaction Effects

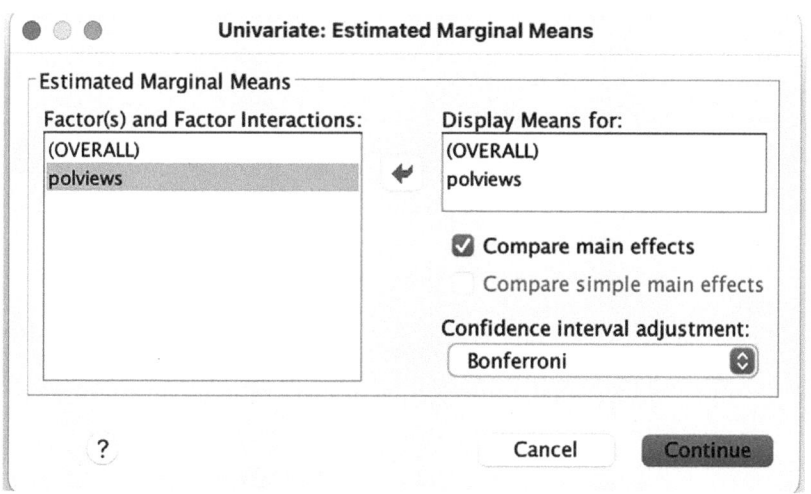

Figure 9.1d Click on Estimated Marginal Means and Transfer Variables to Display Means for Factor and Check off Compare Main Effects and Select Bonferroni

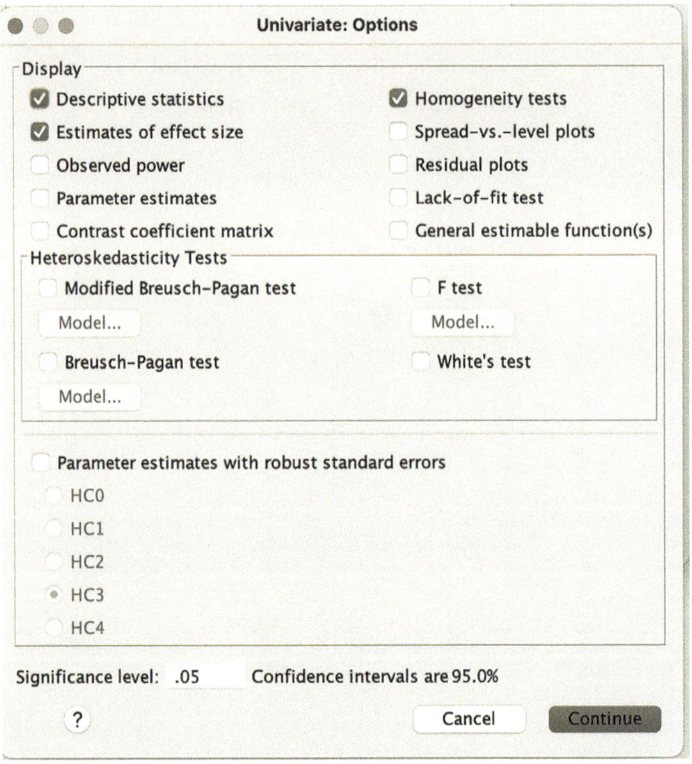

Figure 9.1e Click on Univariate Options, Check off Descriptives, Homogeneity of Variance Test, and Estimates of Effect Size

Assumption Check:

Levene's Test of Equality of Error Variances[a]

Dependent Variable: NUMBER OF CHILDREN

F	df1	df2	Sig.
.290	6	1390	.942

Tests the null hypothesis that the error variance of the dependent variable is equal across groups.

a. Design: Intercept + age + polviews + polviews * age

Tests of Between–Subjects Effects

Dependent Variable: NUMBER OF CHILDREN

Source	Type III Sum of Squares	df	Mean Square	F	Sig.	Partial Eta Squared
Corrected Model	514.817[a]	13	39.601	18.142	<.001	.146
Intercept	.305	1	.305	.140	.709	.000
age	210.656	1	210.656	96.507	<.001	.065
polviews	17.959	6	2.993	1.371	.223	.006
polviews * age	16.566	6	2.761	1.265	.271	.005
Error	3018.807	1383	2.183			
Total	7869.000	1397				
Corrected Total	3533.623	1396				

a. R Squared = .146 (Adjusted R Squared = .138)

Figure SPSS Output #9.1 COMPARISON OF MEANS Analysis of Covariance (ANCOVA)

Between–Subjects Factors

		Value Label	N
THINK OF SELF AS LIBERAL OR CONSERVATIVE	1	EXTREMELY LIBERAL	52
	2	LIBERAL	148
	3	SLIGHTLY LIBERAL	152
	4	MODERATE	590
	5	SLGHTLY CONSERVATIVE	198
	6	CONSERVATIVE	217
	7	EXTRMLY CONSERVATIVE	40

Descriptive Statistics

Dependent Variable: NUMBER OF CHILDREN

THINK OF SELF AS LIBERAL OR CONSERVATIVE	Mean	Std. Deviation	N
EXTREMELY LIBERAL	1.60	1.563	52
LIBERAL	1.55	1.789	148
SLIGHTLY LIBERAL	1.58	1.520	152
MODERATE	1.79	1.577	590
SLGHTLY CONSERVATIVE	1.70	1.599	198
CONSERVATIVE	1.95	1.524	217
EXTRMLY CONSERVATIVE	2.25	1.498	40
Total	1.76	1.591	1397

Levene's Test of Equality of Error Variances[a]

Dependent Variable: NUMBER OF CHILDREN

F	df1	df2	Sig.
.347	6	1390	.912

Tests the null hypothesis that the error variance of the dependent variable is equal across groups.

a. Design: Intercept + age + polviews

Tests of Between–Subjects Effects

Dependent Variable: NUMBER OF CHILDREN

Source	Type III Sum of Squares	df	Mean Square	F	Sig.	Partial Eta Squared
Corrected Model	498.250[a]	7	71.179	32.572	<.001	.141
Intercept	10.079	1	10.079	4.612	.032	.003
age	466.582	1	466.582	213.510	<.001	.133
polviews	13.697	6	2.283	1.045	.394	.004
Error	3035.373	1389	2.185			
Total	7869.000	1397				
Corrected Total	3533.623	1396				

a. R Squared = .141 (Adjusted R Squared = .137)

Pairwise Comparisons

Dependent Variable: NUMBER OF CHILDREN

(I) THINK OF SELF AS LIBERAL OR CONSERVATIVE	(J) THINK OF SELF AS LIBERAL OR CONSERVATIVE	Mean Difference (I–J)	Std. Error	Sig.[a]	95% Confidence Interval for Difference[a]	
					Lower Bound	Upper Bound
EXTREMELY LIBERAL	LIBERAL	.185	.239	1.000	-.541	.911
	SLIGHTLY LIBERAL	.132	.238	1.000	-.592	.855
	MODERATE	.004	.214	1.000	-.649	.656
	SLGHTLY CONSERVATIVE	.034	.231	1.000	-.668	.736
	CONSERVATIVE	-.077	.229	1.000	-.774	.620
	EXTRMLY CONSERVATIVE	-.355	.312	1.000	-1.303	.594
LIBERAL	EXTREMELY LIBERAL	-.185	.239	1.000	-.911	.541
	SLIGHTLY LIBERAL	-.053	.171	1.000	-.573	.466
	MODERATE	-.181	.136	1.000	-.595	.233
	SLGHTLY CONSERVATIVE	-.151	.161	1.000	-.640	.338
	CONSERVATIVE	-.262	.158	1.000	-.742	.219
	EXTRMLY CONSERVATIVE	-.539	.264	.860	-1.342	.263
SLIGHTLY LIBERAL	EXTREMELY LIBERAL	-.132	.238	1.000	-.855	.592
	LIBERAL	.053	.171	1.000	-.466	.573
	MODERATE	-.128	.135	1.000	-.538	.282
	SLGHTLY CONSERVATIVE	-.098	.159	1.000	-.583	.387
	CONSERVATIVE	-.208	.157	1.000	-.686	.269
	EXTRMLY CONSERVATIVE	-.486	.263	1.000	-1.287	.314
MODERATE	EXTREMELY LIBERAL	-.004	.214	1.000	-.656	.649
	LIBERAL	.181	.136	1.000	-.233	.595
	SLIGHTLY LIBERAL	.128	.135	1.000	-.282	.538
	SLGHTLY CONSERVATIVE	.030	.121	1.000	-.339	.400
	CONSERVATIVE	-.080	.117	1.000	-.438	.277
	EXTRMLY CONSERVATIVE	-.358	.242	1.000	-1.094	.377
SLGHTLY CONSERVATIVE	EXTREMELY LIBERAL	-.034	.231	1.000	-.736	.668
	LIBERAL	.151	.161	1.000	-.338	.640
	SLIGHTLY LIBERAL	.098	.159	1.000	-.387	.583
	MODERATE	-.030	.121	1.000	-.400	.339
	CONSERVATIVE	-.111	.146	1.000	-.554	.333
	EXTRMLY CONSERVATIVE	-.388	.256	1.000	-1.169	.392
CONSERVATIVE	EXTREMELY LIBERAL	.077	.229	1.000	-.620	.774
	LIBERAL	.262	.158	1.000	-.219	.742
	SLIGHTLY LIBERAL	.208	.157	1.000	-.269	.686
	MODERATE	.080	.117	1.000	-.277	.438
	SLGHTLY CONSERVATIVE	.111	.146	1.000	-.333	.554
	EXTRMLY CONSERVATIVE	-.278	.254	1.000	-1.052	.496
EXTRMLY CONSERVATIVE	EXTREMELY LIBERAL	.355	.312	1.000	-.594	1.303
	LIBERAL	.539	.264	.860	-.263	1.342
	SLIGHTLY LIBERAL	.486	.263	1.000	-.314	1.287
	MODERATE	.358	.242	1.000	-.377	1.094
	SLGHTLY CONSERVATIVE	.388	.256	1.000	-.392	1.169
	CONSERVATIVE	.278	.254	1.000	-.496	1.052

Based on estimated marginal means

a. Adjustment for multiple comparisons: Bonferroni.

Estimated Marginal Means

THINK OF SELF AS LIBERAL OR CONSERVATIVE

Estimates

Dependent Variable: NUMBER OF CHILDREN

THINK OF SELF AS LIBERAL OR CONSERVATIVE	Mean	Std. Error	95% Confidence Interval	
			Lower Bound	Upper Bound
EXTREMELY LIBERAL	1.780[a]	.205	1.377	2.183
LIBERAL	1.595[a]	.122	1.356	1.833
SLIGHTLY LIBERAL	1.648[a]	.120	1.413	1.884
MODERATE	1.776[a]	.061	1.657	1.896
SLGHTLY CONSERVATIVE	1.746[a]	.105	1.540	1.952
CONSERVATIVE	1.857[a]	.101	1.659	2.054
EXTRMLY CONSERVATIVE	2.134[a]	.234	1.676	2.593

a. Covariates appearing in the model are evaluated at the following values:
AGE OF RESPONDENT = 45.69.

SPSS Example: Technical and Substantive Interpretation

Comparison of Means Hypothesis Testing: ANCOVA (Analysis of Covariance)

Research Question: Are there significant mean differences for the DV, number of children, in the main effects of political views, after controlling for age of respondent?

Null Hypothesis: There are no statistically significant average or mean differences in number of children, in the main effects of political views, after controlling for age of respondent.

Research Hypothesis: There are statistically significant average or mean differences in number of children, in the main effects of political views, after controlling for age of respondent.

Technical interpretation

Between-subjects factors and descriptive statistics: The overall sample size is 1397, in which 52, 148, and 152 identify as some range of liberal, from extremely liberal to slightly liberal; 590 are deemed moderate; 198, 217, 40 identify as belonging to a range of conservatism from slight to extremely, with standard deviations ranging from 1.498 to 1.789. The average number of children, on average, ranges from 1.60 to 2.25. Most, regardless of political views, are having about 2 children, with 2.25 being the highest. The adjusted estimated means also show a similar pattern.

Levene's Test for Equality of Variances & Homogeneity of Regression Slopes: Levene's test for Equality of Variances suggests a statistically significant outcome (F = 0.347, 0.912, $p > 0.05$) and therefore, equal variances are assumed, and the assumption is met. The interaction effect clearly demonstrates that there is no significant interaction effect and the assumption of homogeneity of regression slopes is met (F = 1.265, $p > 0.05$).

Test statistic, Effect size and Post-hoc test: The main effect (A), political views, being extremely liberal to extremely conservative with respect to number of children, is not statistically significant (F = 1.371, $p > 0.05$) suggestive of non-significant average group differences. The means are somewhat similar. However, the covariate of age indicates statistical significance (F = 96.507, $p < 0.05$). The covariate influenced the DV, number of children significantly. There is partial rejection of the null hypothesis and thus only the covariate, age, influenced the outcome, number of children. Political views, identifying as liberal or conservative, did not make a significant difference. The effect size, partial eta squared indicates that the main effects of age are highest at 13.3% explained variance, followed by political views at 0.4% explained variance in number of children. The Adjusted R^2 or coefficient of determination of the overall model for the factor or IV, political views and covariate, age, explained about 13.7% of the variance in the DV, number of children. About 86.3% remains unexplained. The post-hoc test showed no significant average differences amongst the response attributes. Thus, the average differences were not large enough to report.

Substantive Interpretation

Overall, this statistical test, ANCOVA, showed the power of a covariate like age of respondent and how it influences # of children. The covariate superseded the factor, political views. While the means and adjusted means did not show much variation, on average most liberals, moderate, and conservatives settled on average on two children. Identifying as a particular political party did not show any statistically significant average differences. The average differences were near none. Age, the covariate variable, however, did influence the DV # of children. This clearly indicated the age of respondent brought about a spurious effect on number of children and thus political views must be analyzed in combination with the covariate of age. Age combined with political views becomes important to understand.

Final Thoughts

This comparison of means hypothesis test is a much more advanced test and most often discussed and utilized in complex and advance research design and classes. It combines both ANOVA and regression simultaneously. The goal of ANCOVA, as discussed, is to examine how one or more categorical IV or factor influences an interval-ratio continuous DV, while controlling for a specific covariate. The addition of the covariate is the beauty of ANCOVA as it really assesses group averaged differences while controlling for potential effects of a third unseen variable. ANCOVAs explanatory power of the model is great and much better than a simple ANOVA model, even though it has many assumptions tagged along with it. Violations of any assumption impacts the reliability and validity of the test, as well as generalizability. The use of control variables in the social sciences is grand. ANCOVA is one technique that allows for further exploration of control variables or covariates in establishing our conclusions to any hypothesis. Overall, ANCOVA is a statistical test that ensures that no unseen third variable is influencing your model(s). It ensures clarity and understanding of the hypothesis being tested. If all components are present to run an ANCOVA it is well worth a researcher's time as its power of explanation is much stronger, compared to the other tests discussed.

Keywords and Definitions

Analysis of Covariance or ANCOVA	This is an advanced multivariate ANOVA technique with more than two groups' averages being compared on a particular DV. It is a statistical analysis that combines components of ANOVA and regression. It basically evaluates whether population means of a DV are equal or the same across different levels of a categorical IV, while controlling for the effects of a continuous interval-ratio variable-covariates.
Covariate	A continuous interval-ratio variable or its equivalent (i.e., scaled) that controls for the effects on the DV. Its role is to minimize the effects of variability in the DV due to the covariate. The covariate and the DV should assume a linear and consistent relationship.
Homogeneity of Variances or Levene's Test for Equality of Variances	This ensures equal or not equal variances assumed, where a preference is given to 'equal' variances that variances are equal across all levels of the IV, after accounting for effects of covariates. If the F-statistic is significant we report not equal variances; if the F-statistic is not significant we report unequal or not equal variances. The goal here is to ensure that variability in the DV within each group or level is the same.
Slopes of regression lines	Parallel or homogeneous lines (slopes from same group); there should be no interaction between the IV or factor and the covariate; fail to reject the null hypothesis and thus should not be statistically significant. Build custom model, then run full model. The effect of the covariate(s) on the DV should be similar across groups.
Effect size	The magnitude of effect size in ANCOVA measures the true significance of observed differences is eta-squared for each independent factor(s) and the overall model.

Eta^2

The variance is explained in the DV by the factor(s) or main effect(s) and covariate(s) together. The closer this number is to 1, the larger the effect size. The closer to 0, the smaller the effect size.

Adjusted R^2

This also provides overall effect of the entire model out of 100%.

Test Your Knowledge

1 _____ is a hypothesis testing statistical analysis that adds a 'covariate' to the model.

   a   ANOVA
   b   MANOVA
   c   MANCOVA
   d   ANCOVA
   e   Two-way factorial ANOVA

2 In ANCOVA, the _____ is interval-ratio continuous and controls for the effects of the DV.

   a   Concomitant
   b   Covariate
   c   Control
   d   IV
   e   a, b, c

3 The assumptions of ANCOVA are many, however, two key assumptions that if violated impact the findings and generalizability of data.

   a   Slopes of regression lines
   b   Homogeneity of variance
   c   Data
   d   Confidence level
   e   a and b

4 Equal variances are a must in ANCOVA and occur when:

   a $p < 0.05$
   b $p > 0.05$
   c   Fail to reject the null hypothesis
   d   a and c
   e   a and b

5 ANCOVA is a statistical test that is a combination of:

   a   ANOVA and regression
   b   Two-way ANOVA and correlation
   c   Regression and MANOVA
   d   ANOVA and Two-way ANOVA
   e   None of the above

6 Eta^2 and Adjusted R^2 estimate effect size in ANCOVA, what is the difference?

    a Eta^2 examines independent effects and Adjust R^2 examines overall variability in the model out of 100%

    b They both assess average differences

    c Adjusted R^2 examines independent effects and Eta^2 examines overall variability in the model out of 100%

    d They are both the same

    e None of the above

7 If assumptions are violated then what is most impacted in ANCOVA?

    a Generalizability of data

    b Reliability and validity of data

    c Nothing

    d The calculations are all wrong

    e a and b

8 The order of least to most complex and power of explanation for the ANOVA family is:

    a One-way ANOVA, Two-way ANOVA, and ANCOVA

    b ANCOVA, One-way ANOVA, Two-way ANOVA

    c Two-way ANOVA and ANCOVA

    d ANCOVA only

    e ANOVA family is a powerless family with many Type 1 and Type 2 errors

9 The ANOVA family is an advanced multivariate ANOVA technique with more than two groups' averages being compared on a particular DV. It is a statistical analysis that combines components of ANOVA and regression. It basically evaluates whether population means of a DV are equal or the same across different levels of a categorical IV, while controlling for the effects of a continuous interval-ratio variable-covariates.

    a True

    b False

10 Homogeneity of variances are parallel or homogeneous (slopes from same group); there should be no interaction between the IV or factor and the covariate; fail to reject the null hypothesis and thus should not be statistically significant. Build custom model, then run full model. The effect of the covariate(s) on the DV should be similar across groups.

    a True

    b False

Multivariate Hypothesis Testing Using Comparison of Means using MANOVA & MANCOVA

Often when investigating the social world our curiosity goes beyond a single particular outcome and there is a broader vision at work that may entail other outcomes or social phenomenon. For example, we may be interested in the psychological scales of self-esteem and anxiety scores amongst students to examine average differences and its effect on the race of the respondent. For example, this could examine the difference in the multivariate means of self-esteem and anxiety scores between four racial groups – Whites, South Asian, Blacks, Other. At least one of the DVs, anxiety or self-esteem, will have a difference in their multivariate means. Consequently, when we test and compare average differences there may be an interest or interest in, perhaps, more than one DV or 'sets' of DVs. So far, all the statistical tests discussed up to this point only focus on one DV.

The comparison of means hypothesis testing techniques discussed in this part of the book is MANOVA, which is simply an extension of the univariate analysis of variance. In analysis of variance, we examine the one dependent variable with the grouping independent variable or independent variables or factors. In MANOVA, the analysis extends to multiple DVs. Thus, there are multiple measurable social phenomena that can be tested through hypothesis testing. Through examining patterns of average group differences, MANOVA provides great insights into increase complex relationships. Its core uniqueness lies in its ability to work with multiple DVs and understand how different groups or conditions diverge across a set of dependent variables or outcomes. The comparison of means multivariate test, MANOVA or Multivariate Analysis of Variance (MANOVA) is a statistical test that allows to concurrently analyze two or more DVs. This type of statistical analysis is most beneficial when the combined effects of factors or IVs becomes critical to a specific set of correlated DVs. MANOVA creates a holistic analysis of the relationships in question and makes efficient use of data. Moreover, due to its efficiency, MANOVA reduces the occurrence of Type 1 errors, thus in turn improving the reliability of data. The increased statistical power of the test speaks for itself and is especially increased when there are decent correlations amongst the DVs or outcome variables. The ability to assess average or mean differences on the set of DVs in its entirety makes this advanced statistical technique distinctive.

Main Differences in ANCOVA vs. MANOVA vs. MANCOVA

Main difference	ANCOVA	MANOVA	MANCOVA
Inclusion of covariate	Adds covariate or control in the model	No covariate in the model	Adds covariate or control in the model
Objective	Reports on significant average differences on a DV, while controlling for covariates; adjust means accordingly	Reports on significant average differences on a 'set' of DVs with no controls or covariates present	Reports on significant average differences on a 'set' of DVs with controls or covariates present
Handling differences	Covariates play a pivotal role in adjusting for any baseline differences	Covariates do not play any role in adjusting for any baseline differences	Covariates play a role in adjusting for any baseline differences
Statistical power	Increased statistical power with covariate	Comprehensive explanations with 'sets' of DVs and at least two outcomes	Comprehensive explanations with 'sets' of DVs and at least two outcomes with covariate(s)
Homogeneity of regression slopes	The relationship between the DV and covariate is similar across all levels of the IV(s) or factor(s)	The variances amongst the 'sets' of DVs is equal across all groups tested	The relationship between the 'sets' of DV and covariate is equal across all groups tested

The conceptualization of the DVs should be clear and make sense for making it empirically meaningful. While having multiple DVs may make the statistical analysis more complex, there are two key reasons why it may be necessary for the analysis:

1   It allows for a comprehensive picture of the social phenomenon being tested.
2   Any treatment may impact individuals in varying ways.

MANOVA, like any other inference based statistical test, has many assumptions that must be fulfilled or satisfied prior to making adequate generalizations from a sample to a population. Additionally, ensuring that assumptions are met is important to avoid reporting inaccurate results. If there is any assumption violation, then it must be reported and discussed as a limitation of the data or nature of data. It is always important to report back whether assumptions were met as they make a difference with respect to generalizability of outcomes and findings.

Assumptions	Explanation
1. Data or level of measurement of variables	MANOVA assumes that the independent variables are categorical in nature and the dependent variables are continuous interval-ratio or scaled variables
2. Sampling method and sample size	Random sampling and large sample
3. Independence of observations	Data observations must be independent of each other; no influence by other observations
4. Shape of the distribution	Normally distributed DVs
5. Homogeneity of variances	The variances of the DV are equal across all groups; Levene's Test or Box's M assess homogeneity of variances
6. Absence of multicollinearity	The IVs are not highly correlated with each other; each IV brings its own unique variance
7. Linearity	The relationship between the IV and DV is linear
8. Slopes of regression lines	The slopes of the regression lines are parallel or homogeneous (slopes from same group); fail to reject the null hypothesis and thus should not be statistically significant
9. Confidence level	The amount of risk you are willing to take: 95%, 0.05 alpha
10. Sample statistic	Multivariate F-ratio or F-statistic; rejects the H_o or fails to reject it
11. Outliers	Minimize outliers or extreme scores

In ANOVA, we compare grouping independent variables with one dependent variable, but in MANOVA, we compare 'sets' of at least two dependent variables with the grouping variable, factor or IV. The key questions MANOVA tries to understand, and answer are:

1   What are the main effects of the independent variables?
2   What are the interactions among the independent variables?
3   What is the importance of the dependent variables and are there significant average differences in the groups of the DV?
4   What is the strength of association between dependent variables?

Further investigation may add a 'covariate' to the model creating a Multivariate Analysis of Covariance Design (MANCOVA) statistical design. The key difference between MANOVA and MANCOVA is the incorporation of 'covariates' or control variables to the model. By controlling for the effects of unseen variables, the accuracy of the analysis is much improved. Just like in ANCOVA models, MANCOVA consider if average differences exist once the model has been adjusted for by a covariate(s). Calculations of MANOVA and MANCOVA are based on matrix algebra. In MANOVA and MANCOVA models, the DVs or outcome variables should have some degree of linearity and share theoretical and logical meaning. What are the effects of covariates and if average group differences remain after adjusting for a covariate(s)? Like MANOVA, MANCOVA is bound to assumptions as well. These assumptions must not be violated to maximize unbiased statistical outcomes. Most assumptions are tied to the covariate in MANCOVA.

Assumptions	Explanation
1. Data or level of measurement of variables	MANCOVA assumes that the independent variables are categorical in nature and the dependent variables, as well as covariate(s) are continuous interval-ratio or scaled variables
2. Sampling method and sample size	Random sampling and large sample > than the DVs
3. Independence of observations	Data observations must be independent of each other; no influence by other observations
4. Shape of the distribution	Normally distributed DVs
5. Homogeneity of variances	The variances of the DVs and covariate(s) are equal across all groups; Levene's Test or Box's M assess homogeneity of variances if this is less than 0.05 you have violated variance assumption; anything greater than 0.05 indicates variances are equal and no violation has occurred; Levene's test should be not significant as well, $p > 0.05$, so equal variances generated
6. Modest multicollinearity	DVs are moderately correlated and not too high; run a correlation check to ensure they are not highly correlated
7. Linearity	Having a straight-line relationship between each set of DVs using matrix of scatterplots
8. Confidence level	The amount of risk you are willing to take: 95%, 0.05 alpha
9. Sample statistic	Multivariate tests: Wilks' Lambda, Hotelling's Race, Pillai's Trace indicate statistically significant differences amongst the linear combination of the DVs; rejects the H_o or fails to reject it
10. Outliers	Minimize outliers or extreme scores

In MANOVA and MANCOVA, the research question must include the factors or IV(s), the 'sets' of DV and the covariate variable. The research question and the null and research hypothesis are written in specific terms. Let's see the example below:

Research Question: Are there significant mean differences in the combined DVs, current salary and education for minority classification (IV_a), while controlling for beginning salary?

Null hypothesis: There are no significant mean differences in the combined 'sets' of DVs between the groups of Factor A, while controlling for the covariate?

There are no statistically significant average or mean differences in the combined DV's, current salary and education for minority classification (IV_a), while controlling for the covariate, beginning salary?

Research hypothesis: There are significant mean differences in the combined 'sets' of DVs between the groups of Factor A, while controlling for a covariate?

There are statistically significant average or mean differences in the combined DV's, current salary and education for minority classification (IV_a), while controlling for the covariate, beginning salary?

If there is no covariate, we just do not include it.

For MANOVA and MANCOVA the variables needed are:

- Independent variable: Requires two or more nominal or ordinal categorical IVs with two levels or more or response attributes
- Dependent variable: Two multiple interval-ratio continuous or scaled variables with a shared conceptual meaning
- Covariate(s): 1–2 Interval-ratio continuous or scaled variables with a significant association with the DV

For example, teaching methods, gender, treatment groups, etc., can assume the role of the IV and the DV can a scale or a score or years of education. Sometimes, additional control variables, may be added to the MANOVA model. Hence, it becomes a Multivariate Analysis of Covariance model or simply put as MANCOVA. To test for spurious effects, the control variable assesses the model.

- Each makes a 'new' unique contribution to the depth of analysis by adding a covariate to the model and making use of several DVs.
- What statistical test you choose depends on your research question, variables, and what and how you would like to explain you research objectives

Effect Size and MANOVA and MANCOVA

The magnitude of effect size in MANOVA and MANCOVA is to measure the true significance of observed differences is eta^2 for each independent factor(s) and the overall model. Again, like the other ANOVA family tests, eta^2 represents the variance explained in the DV by the factor(s) or main effect(s) and covariate(s) together. The closer this number is to 1, the larger the effect size. The closer to 0, the smaller the effect size. The Adjusted R^2 also provides overall effect of the entire model out of 100%. Again, this is an important measure of the strength of an effect and tells the proportion of explanation accounted in the DV for each factor(s) after accounting for the effects of the covariate(s).

MANOVA and MANCOVA and Post-hoc Testing

In MANOVA and MANCOVA, as like other ANOVA testing, the F-statistic sample statistic, only highlights an average or mean difference, but does not inform us of which groups significantly differ or not, provided there are more than two groups for each factor or IV. If there are not more than two groups, a post-hoc test will not be performed and that is okay. To establish precision in mean differences, a post-hoc test needs to be run to establish exactly where the average group differences lie. This is important to know. Knowing that there are mean differences is great, but to complete the analysis reporting of which groups significantly differ is equally important to this type of analysis. There are many post-hoc testing procedures based on the research design and nature of data, like Tukey, Scheffé, Dunnett's, and Bonferroni, that the ANOVA family offers. Bonferroni post-hoc test is the most robust one and popular one that is generally used. Post-hoc testing is an integral part of the MANOVA test and must be discussed for a complete analysis of results.

Activity Alert

How is MANOVA different then ANCOVA?
What are the levels of measurement required to run this test?
How does MANOVA differ from MANCOVA?
What are the assumptions of the MANOVA test?

Key Statistics to Report

1 Descriptive statistics: Report basic descriptive statistics, sample size (N), averages, grand mean, standard deviation, minimum and maximum values; provide trends and patterns of data; this provides readers with a quick overview of each group characteristics as it relates to the DV or social phenomenon by each factor.
2 Box's M Test and Levene's Test for Equality of Variances: Report whether equal or not equal variances are assumed; Equal ($p > 0.05$) or not equal variances ($p < 0.05$): Tests the assumption of equal or not equal variances, remember you can never have two nots together (i.e., not significant, means equal variances; significant means not equal variances); non-violation of the assumption yields equal variances amongst groups. Equal variances are ideal.
3 Multivariate Tests: Wilks' Lambda is used to the sample statistic used to determine statistical significance when equal variances are assumed, and no violations have occurred. This is the most favorable statistical significance test used; however, in case of violations resort to Pillai Trace which is robust. Report the calculated test or sample statistic that evaluates statistical significance to reject or fail to reject or partial rejection of the null hypothesis of overall comparison of means; This value reports significant mean differences or not; Degrees of freedom can be reported but is not mandatory.
4 Between-subject effects: If a significant outcome is reached with multivariate tests, then further analysis is done on the DVs and factors to see if each factor differs on all DVs? If it is statistically significant, then differences with respect to that factor are present.
5 Partial eta^2 (effect size) & Adjusted R^2: magnitude of effect size and reports the proportion of total variance attributed to group differences and whether a meaningful difference is apparent for each factor independent of each other and the overall full model using Adjusted R^2. R^2 can be reported, it is the less conservative number.
6 Post-hoc testing: Report specific significant average or mean group differences. Bonferroni is the most robust test. Other post-hoc test may also be run.
7 Always tell the reader, what the confidence level and alpha is set to. For example, is it at 95%, alpha of 0.05 or 99%, alpha of 0.001?

MANOVA in SPSS

1 Click on Analyze > General Linear Model > Multivariate
2 Move the variables of interest into the Dependent List (DVs) and Fixed Factors List (IVs)
3 Check Options > Click on EM Means >Transfer variable from the Factor(s) and Factor Interactions box to the Display means for Factor and check off compare main effects and select Bonferroni

4   Check off Descriptives, Homogeneity of Variance Test, and Estimates of Effect Size
5   OK to generate an MANOVA

MANCOVA in SPSS

1   Click on Analyze > General Linear Model > Multivariate
2   Move the variables of interest into the Dependent List (DVs), Fixed Factors List (IVs), and Covariate Box
3   Check Options > Click on EM Means >Transfer variable from the Factor(s) and Factor Interactions box to the Display means for Factor and check off compare main effects and select Bonferroni
4   Check off Descriptives, Homogeneity of Variance Test, and Estimates of Effect Size
5   OK to generate an MANCOVA

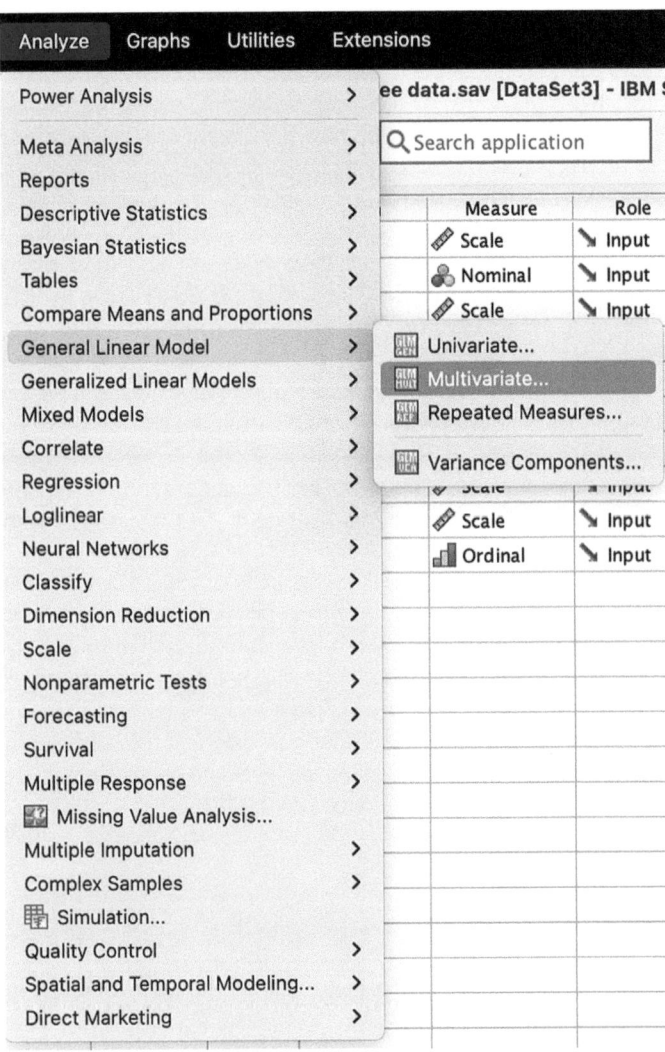

Figure 9.2a SPSS Command for MANOVA

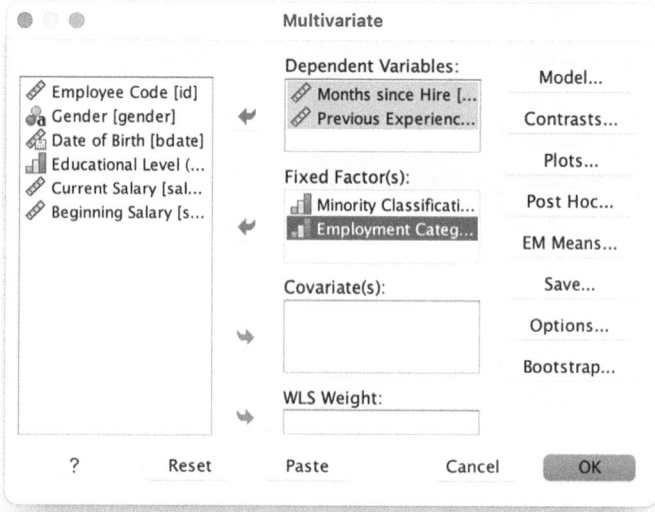

Figure 9.2b Move the variables of interest into the Dependent List (DVs) and Fixed Factors List (IVs)

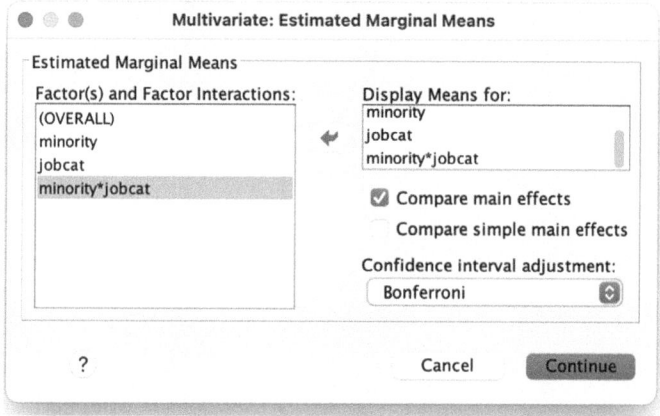

Figure 9.2c Check Options and Click on EM Means, Transfer variable from the Factor(s) and Factor Interactions box to the Display means for Factor and check off compare main effects and select Bonferroni

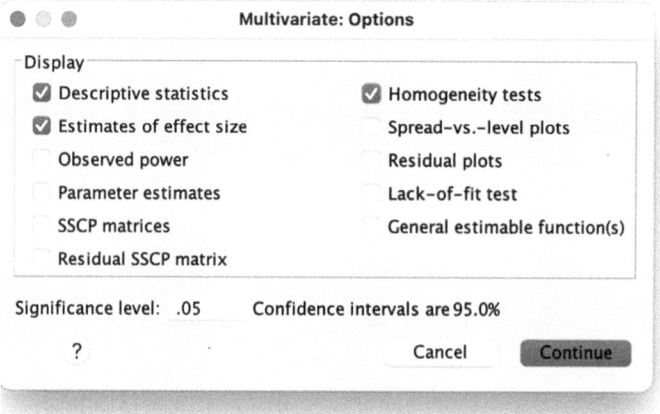

Figure 9.2d Check off Descriptives, Homogeneity of Variance Test, and Estimates of Effect Size

Between–Subjects Factors

		Value Label	N
Employment Category	1	Clerical	363
	2	Custodial	27
	3	Manager	84
Minority Classification	0	No	370
	1	Yes	104

Descriptive Statistics

	Employment Category	Minority Classification	Mean	Std. Deviation	N
Months since Hire	Clerical	No	80.58	10.136	276
		Yes	82.60	9.931	87
		Total	81.07	10.110	363
	Custodial	No	80.86	7.873	14
		Yes	82.31	9.366	13
		Total	81.56	8.487	27
	Manager	No	81.75	10.288	80
		Yes	69.25	3.594	4
		Total	81.15	10.410	84
	Total	No	80.85	10.082	370
		Yes	82.05	9.978	104
		Total	81.11	10.061	474
Previous Experience (months)	Clerical	No	78.32	95.695	276
		Yes	106.34	91.226	87
		Total	85.04	95.275	363
	Custodial	No	352.21	85.029	14
		Yes	239.85	85.815	13
		Total	298.11	101.426	27
	Manager	No	74.43	71.490	80
		Yes	141.50	90.497	4
		Total	77.62	73.260	84
	Total	No	87.84	104.557	370
		Yes	124.38	100.070	104
		Total	95.86	104.586	474

Box's Test of Equality of Covariance Matrices[a]

Box's M	19.716
F	1.203
df1	15
df2	1553.632
Sig.	.262

Tests the null hypothesis that the observed covariance matrices of the dependent variables are equal across groups.

Figure SPSS Output #9.2 Multivariate Analysis of Variance (MANOVA)

Multivariate Tests[a]

Effect		Value	F	Hypothesis df	Error df	Sig.	Partial Eta Squared
Intercept	Pillai's Trace	.923	2799.768[b]	2.000	467.000	<.001	.923
	Wilks' Lambda	.077	2799.768[b]	2.000	467.000	<.001	.923
	Hotelling's Trace	11.990	2799.768[b]	2.000	467.000	<.001	.923
	Roy's Largest Root	11.990	2799.768[b]	2.000	467.000	<.001	.923
jobcat	Pillai's Trace	.220	28.956	4.000	936.000	<.001	.110
	Wilks' Lambda	.782	30.527[b]	4.000	934.000	<.001	.116
	Hotelling's Trace	.276	32.101	4.000	932.000	<.001	.121
	Roy's Largest Root	.264	61.804[c]	2.000	468.000	<.001	.209
minority	Pillai's Trace	.004	.995[b]	2.000	467.000	.370	.004
	Wilks' Lambda	.996	.995[b]	2.000	467.000	.370	.004
	Hotelling's Trace	.004	.995[b]	2.000	467.000	.370	.004
	Roy's Largest Root	.004	.995[b]	2.000	467.000	.370	.004
jobcat * minority	Pillai's Trace	.049	5.832	4.000	936.000	<.001	.024
	Wilks' Lambda	.952	5.831[b]	4.000	934.000	<.001	.024
	Hotelling's Trace	.050	5.831	4.000	932.000	<.001	.024
	Roy's Largest Root	.035	8.268[c]	2.000	468.000	<.001	.034

a. Design: Intercept + jobcat + minority + jobcat * minority

b. Exact statistic

c. The statistic is an upper bound on F that yields a lower bound on the significance level.

Tests of Between-Subjects Effects

Source	Dependent Variable	Type III Sum of Squares	df	Mean Square	F	Sig.	Partial Eta Squared
Corrected Model	Months since Hire	884.059[a]	5	176.812	1.761	.119	.018
	Previous Experience (months)	1329102.25[b]	5	265820.451	32.357	<.001	.257
Intercept	Months since Hire	534919.281	1	534919.281	5327.084	<.001	.919
	Previous Experience (months)	2313218.185	1	2313218.185	281.578	<.001	.376
jobcat	Months since Hire	537.042	2	268.521	2.674	.070	.011
	Previous Experience (months)	1015443.050	2	507721.525	61.803	<.001	.209
minority	Months since Hire	191.640	1	191.640	1.908	.168	.004
	Previous Experience (months)	700.237	1	700.237	.085	.770	.000
jobcat * minority	Months since Hire	759.154	2	379.577	3.780	.024	.016
	Previous Experience (months)	130372.943	2	65186.472	7.935	<.001	.033
Error	Months since Hire	46994.236	468	100.415			
	Previous Experience (months)	3844704.555	468	8215.181			
Total	Months since Hire	3166222.000	474				
	Previous Experience (months)	9529528.000	474				
Corrected Total	Months since Hire	47878.295	473				
	Previous Experience (months)	5173806.810	473				

a. R Squared = .018 (Adjusted R Squared = .008)

b. R Squared = .257 (Adjusted R Squared = .249)

Pairwise Comparisons

Dependent Variable	(I) Employment Category	(J) Employment Category	Mean Difference (I–J)	Std. Error	Sig.[b]	95% Confidence Interval for Difference[b] Lower Bound	Upper Bound
Months since Hire	Clerical	Custodial	.008	2.026	1.000	−4.859	4.875
		Manager	6.091	2.640	.064	−.252	12.433
	Custodial	Clerical	−.008	2.026	1.000	−4.875	4.859
		Manager	6.082	3.212	.177	−1.634	13.798
	Manager	Clerical	−6.091	2.640	.064	−12.433	.252
		Custodial	−6.082	3.212	.177	−13.798	1.634
Previous Experience (months)	Clerical	Custodial	−203.697*	18.323	<.001	−247.720	−159.673
		Manager	−15.629	23.878	1.000	−72.999	41.741
	Custodial	Clerical	203.697*	18.323	<.001	159.673	247.720
		Manager	188.068*	29.048	<.001	118.276	257.860
	Manager	Clerical	15.629	23.878	1.000	−41.741	72.999
		Custodial	−188.068*	29.048	<.001	−257.860	−118.276

Based on estimated marginal means

*. The mean difference is significant at the .05 level.

b. Adjustment for multiple comparisons: Bonferroni.

SPSS Example: Technical and Substantive Interpretation

Comparison of Means Hypothesis Testing: A Two-Way Between Groups MANOVA

Research Question: Are there significant mean differences in the combined DVs, months since hire and previous experience, for factors, minority classification and job category?

Null Hypothesis: There are no statistically significant mean differences in the combined DVs, months since hire and previous experience, for factors, minority classification and job category?

Research Hypothesis: There are statistically significant mean differences in the combined DVs, months since hire and previous experience, for factors, minority classification and job category?

Technical interpretation

Between-subjects factors and Descriptive statistics: The overall sample size is 948, in which 363, 27, and 84 identify as clerical, custodial, and managers, with 370 identifying as non-minorities and 104 identifying as minorities. The average months since hire for clerks and custodians depicted a similar pattern, ranging from about 80 months to 82 months, with non-minorities being newer on the job. In the case of managers, on average minorities' month since hire was 69.25 months, compared to 81.75 months for non-minorities. Thus, non-minorities, on average, were on the job for a longer period. There were mean differences seen, especially in the managerial category, compared to clerks and custodians. The second DV, previous experience (in months), on average, for clerks and custodians, was less for clerks and greater for custodians, with minority clerks having more months of previous experience and Caucasian custodians having, on average, increased previous experience. Caucasian managers had less previous experience, compared to their minority counterparts with, on average 141.50 months of previous experience, compared to only 74.43 months for Caucasian managers. Custodians across the board, on average, displayed the most months of previous experience, followed by managers and clerks. Thus, the means displayed slight variations in their average differences across the DVs.

Box's Test of Equality of Covariance Matrices and Levene's Test for Equality of Variances: Box's Test of Equality of Covariance Matrices suggests a not significant outcome (F = 1.203, 0.262, $p > 0.05$), as well as the Levene's test for Equality of Variances for both DV's month since hire and previous experience suggest a not statistically significant outcome (F = 1.520, 0.182, $p > 0.05$; F = 0.769, 0.572, $p > 0.05$) and therefore, equal variances are assumed, and the assumption of homogeneity of variances is met.

Multivariate tests, Between subjects effects, Effect size & Post-hoc test: The Wilks' Lambda suggested that were statistically significant differences for job category (clerk, custodian, and manager) on the combined DV's, months since hire and previous experience, F = 30.527, $p < 0.001$. Nevertheless, the Wilks' Lambda for minority classification suggested a non-significant outcome, F = 0.995, 0.370, $p > 0.05$. However, the combined effects of the factors or interaction term, job category*minority classification, indicated a statistical significance outcome (F = 5.832, $p < 0.001$). The interaction term supercedes the main effect and both factors need to be investigated in combination and not separately. When the results for the DV were considered independently, the only difference to reach statistical significance was job category with previous experience, with a partial eta^2 of 20.9% and the combined effects or interaction term of job category with minority classification with previous experience and months since hire with a partial eta^2 of 1.6% and 3.3% respectively. Month since hire did not create any statistically significant differences worth reporting for job category and minority classification as main effects. The Adjusted R^2 for months since hire was 0.8% and about 25% for previous experience. The post-hoc test for job category revealed that clerks and custodians previous experience significant differed by 203 months and manager and custodians by 188 months. The DV, months since hire showed no statistically significant average differences to report. The averages had less variation.

Substantive interpretation

Overall, this statistical test, Two-way between groups MANOVA, showed how multiple DVs, like months since hire and previous experience, display statistically significant average differences for the factors, minority classification and job category. This MANOVA clearly showed that average differences in the DVs was seen when minority classification and job category were combined. Their interaction accounted for average differences. The DV previous experience was at a slight advantage, compared to months since hire. This analysis demonstrated that previous experience is a better contender when both factors are examined together with regards to average differences. Months since hire displayed low variation amongst the groups being tested and thus no significant average differences.

Final Thoughts

MANOVA is an advanced statistical data analysis test that investigates average or mean differences among groups when there are multiple DVs present to test against. Multiple social phenomenon or social outcomes are prudent to certain analyses or questions we may have about the social world, and it becomes important to understand the nuances of such relationships and their factors or IVs that may influence them. MANOVA, like ANCOVA, is an extension of the simple ANOVA model. When two interval-ratio continuous DVs are present, MANOVA analyzes the 'set' of DVs against a nominal or ordinal factor or factor(s) with more than two groups and provides a combined test of significance for all DVs. Selecting MANOVA versus engaging in multiple ANOVA models, reduces error and bias.

MANCOVA, on the other hand, an extension of MANOVA, introduces one or more interval-ratio continuous covariate(s) to the model of 'average differences' on sets of DVs. The key goal is to assess group differences in the multivariate means of dependent variables while controlling for the effects of covariates. This statistical analysis allows to test for average group differences while accounting for the effect of covariates. In summary, MANCOVA provides an all-encompassing analysis of group effects, covariate effects, and their interactions on the dependent variables.

While MANOVA and MANCOVA are powerful advanced statistical tests, they come with their own limitations. First, MANOVA has a running list of assumptions that must be met and not violated. Sometimes, with real data, this becomes difficult to maintain. Additionally, MANOVA is best received when each group has a similar number of observations to avoid biasing results. Ultimately, when it comes to interpreting MANOVA and MANCOVA, the outputs and statistical analysis can be quite convoluted and complex. MANOVAs warrants an advanced statistical research design and thus, requires patience and attention to detail to understand what the findings or results entail. Sometimes, taking a break from the analysis is good and come back to it with fresh eyes.

Keywords and Definitions

MANOVA or Multivariate Analysis of Variance (MANOVA)	A statistical test that allows to concurrently analyze two or more DVs alongside the IVs. Two outcomes are assessed simultaneously.
MANCOVA	An extension of MANOVA introduces one or more interval-ratio continuous covariates to the model of 'average differences' on sets of DVs. The key goal is to assesses group differences in the multivariate means of dependent variables while controlling for the effects of covariates.
Slopes of regression lines	Parallel or homogeneous lines (slopes from same group); there should be no interaction between the IV or factor and the covariate; fail to reject the null hypothesis and thus should not be statistically significant. Build custom model, then run full model. The effect of the covariate(s) on the DV should be similar across groups.
Eta^2	The variance is explained in the DV by the factor(s) or main effect(s) and covariate(s) together. The closer this number is to 1, the larger the effect size. The closer to 0, the smaller the effect size.
Adjusted R^2	This also provides overall effect of the entire model out of 100%. Again, this is an important measure of the strength of an effect and tells the proportion of explanation accounted in the DVs for each factor(s) and after accounting for the effects of the covariate(s).
Post-hoc tests	This needs to be run to establish exactly where the average group differences lie. There are many post-hoc testing procedures based on the research design and nature of data, like Tukey, Scheffé, Dunnett's, and Bonferroni.

Test Your Knowledge

1 _____ is an advanced statistical data analysis test that investigates average or mean differences among groups when there are multiple DVs present to test against.

   a   MANCOVA
   b   MANOVA
   c   Post-hoc
   d   ANOVA
   e   Covariates

2 _____is an extension of MANOVA introduces one or more interval-ratio continuous covariates to the model of 'average differences' on sets of DVs. The key goal is to assesses group differences in the multivariate means of dependent variables while controlling for the effects of covariates.

   a   MANOVA
   b   MANCOVA
   c   Post-hoc
   d   ANOVA
   e   Covariates

3 _____are parallel or homogeneous (slopes from same group); the outcome should be to fail to reject the null hypothesis and thus should not be statistically significant.

   a   Slopes of the regression lines
   b   Multicollinearity
   c   Adjusted R^2
   d   Covariates
   e   Post-hoc

4 What are the assumptions of MANOVA and MANCOVA?

   a   Level of measurement, sampling method, independence of observations, descriptive statistics
   b   Level of measurement, sampling method, independence of observations, shape of distribution, absence of multicollinearity, linearity, slopes of regression lines
   c   No outliers and shape of distribution only
   d   Homoscedasticity and slopes of regression lines
   e   None of the above

5 ____ represents the variance explained in the DV by the factor(s) or main effect(s) and covariate(s) together. The closer this number is to 1, the larger the effect size. The closer to 0, the smaller the effect size.

   a   Partial eta^2
   b Eta^2
   c   Adjusted R^2
   d   MANOVA
   e   MANCOVA

6   Post-hoc tests need not be run to establish exactly where the average group differ-
    ences lie. There are not many post-hoc testing procedures based on the research
    design and nature of data, like Tukey, Scheffé, Dunnett's, and Bonferroni.

    a   True
    b   False

7   Partial eta^2 also provides overall effect of the entire model out of 100%. Again,
    this is an important measure of the strength of an effect and tells the proportion of
    explanation accounted in the DVs for each factor(s) and after accounting for the
    effects of the covariate(s).

    a   True
    b   False

8   What is similar amongst all the tests of the ANOVA family?

    a   They compare means and test hypotheses
    b   They don't compare means and test hypotheses
    c   They predict and explain
    d   They don't predict and explain
    e   They tell us about associations and predictions

9   The MANOVA and MANCOVA models are:

    a   Advanced ANOVA testing
    b   Multivariate testing
    c   Hypothesis testing
    d   Comparison of means testing, one with two DVs and one with a covariate
    e   All of the above

10 The nature of the IV for MANOVA and MANCOVA are:

    a   An IV that has two or more nominal or ordinal categorical IVs with two levels
        or more or response attributes
    b   At least two IVs that are interval-ratio in nature
    c   An IV that is interval-ratio continuous in nature
    d   a and c
    e   None of the above

Part 3

Inferential Statistics with a Focus on Associations and Predictions Using Correlation and Regression Models

The next and final chapters of this book cover data analysis as it pertains to associations and predictions and predictive probabilities with interval-ratio continuous data or its equivalent. This part of the book is an in-depth discussion of the importance of associations and predictions with sophisticated statistical tests, like correlation and regression. Furthermore, these chapters provide insight into the various types of correlations and regressions that can be used by social scientists, the key statistics they report, and their technical and substantive interpretations. Differences between these statistical tests are highlighted and noted.

10 Correlations with Interval-Ratio Continuous Data

Introduction to Associations and Predictions in Inferential Statistics with Interval-ratio Continuous Data

Alas, the journey of our statistical pathways embarks on a path of associations and predictions. These are fundamental concepts used in data analysis under the branch of inferential statistics. In the social sciences, statistical analysis focused on associations and predictions, or predictability becomes an important pathway to making sense of certain variables and relationships we want to test. Sometimes, the questions we are concerned with posit associations and predictions and go beyond the task of exploring trends and patterns of the data and average or mean differences. In fact, we are genuinely interested in how variables co-vary or correlate or associate and how well do certain IVs predict a certain outcome. For example, it might interest us to understand the association between number of times convicted and years of education or self-esteem scores and weight or GPA and hours spent studying. Correlation and regression analyses are at the heart of this section. Despite their popularity, the flexibility and wide-range application of these statistical techniques are sometimes overlooked and misunderstood. Nonetheless, these data analyses techniques are the soul for interval-ratio continuous IVs and DVs. Together, they formulate the essence of discussions around causal linkages in research. These types of analyses are used across many disciplines, like sociology, criminology, education, psychology, science, and science. These statistical techniques transcend into many areas of everyday life in creative ways. This part of the book focuses on the following statistical techniques: Correlations, Bivariate and multiple linear regression, Hierarchical or Block modelling or Incremental regression, and finally Logistic regression analyses. The discussion of correlation focuses on association between the IVs and DV, while the regression chapters focus on predictability and probability. So, what is the difference between these?

Association	Prediction
Understanding the relationship or association between an IV(s) and DV	Making or forecasting predictions about a relationship using information about variables
Example: # of hours studying and GPA	Example: Knowing the # of hours studying and making a prediction about GPA

These two characteristics of association and prediction most often are paired together. In fact, most regression techniques, provide correlational values as a descriptive statistic when run. Associations mostly determine the strength and direction of a relationship but does not explain causality and does not make predictions. The statistical test of correlations which provide a correlation coefficient and scatterplot analyze associations. Prediction, on the other hand, estimates the value of another variable through predictions using a regression equation. Regression analysis goal is to make accurate predictions and estimate a relationship using known values about one or more variables.

DOI: 10.4324/9781003215691-13

Correlation	Regression
Tells which two variables, IV and DV or X and are related	Regression predicts the value of a dependent variable based on the value of at least one independent variable or predictor
It gives you a correlation coefficient value (r) that ranges between –1 and 1 and can be positive or negative	It gives you an equation that represents the best prediction of the dependent variable given the independent variables or predictors
A correlation between X and Y is the same as the correlation between Y and X	The regression of Y on X is different from the regression of X on Y unless the relationship is perfectly linear with no variation
It does not imply causation and does not assume or predict that one variable causes the other to change	Regression can be used to infer causality under specific conditions if the data and research design support such conclusions.
	The slope of the regression line represents the relationship between the variables.

Key Characteristics of Associations and Prediction Relationship Testing for Interval-ratio Continuous Data

Null hypothesis	A statement of no association or prediction
Alternative hypothesis	A statement of association or prediction
Sample test statistic	The sample statistic calculated from the sample data to assess the null hypothesis; in this case it is 'r' and t-statistic
Significance level and p-value	Tells the probability of rejecting the null hypothesis or calculated sample or test statistic, commonly set at an alpha value of 0.05 (95% confidence level); $p < 0.05$, reject the H_o and have significant average differences
Decision rule	The calculated test or sample statistic results in a decision to reject the H_o or fail to reject the H_o

There is a deep-seated connection between association and prediction, and this should not go unnoticed. Association helps reveal relationships and guides regression models. These models are a product of the existing associations. Either way, association and prediction are two peas in a pod. They both statistically make informed decisions about the analysis and provide meaningful insights from the data. Association and predictions guide decision-making at its best, especially when the IV and DV are interval-ratio continuous variables, the highest level of measurement or its equivalent. These next chapters will focus on both the correlation and regression families and their statistical assumptions and power. With the highest levels of measurement, these statistical techniques become critical to the statistical pathways of data analysis.

Association refers to the simple correlation or covariation of two variables or the association amongst them, not necessarily the cause and effect or causation. Some simple questions or social inquiries we may be interested in learning about as a researcher are:

- # of times arrested and # of convictions
- Hours of study and exam scores
- TV hours watched and years of education
- # of children and hours of sleep

- Gender and # of times assaulted (Women are more likely to be assaulted)
- Race and # of times involved in a hate crime incident (Minorities are more likely to be involved in hate crime incident)

These examples simply assess particular associations about various relationships about the social world around us. All these associations are bivariate observations that can be statistically studied and understood. Another major concept that goes hand in hand with association, is prediction or predictability. Prediction is the heart of the final chapter and inferential statistics. It is defined by using known information to estimate or forecast unknown or known values. Predictions anticipate outcomes, trends, or values based on the data or established patterns. Regression analyses of sorts is the core analysis that analyzes predictions.

Pearson Product Moment Correlation (r)

The Pearson Product Moment Correlation (r) is a bivariate measure of association that is heavily linked to interval-ratio continuous data or its equivalent (i.e., dummy coded to 0 and 1 values, with 1 being the reference group or scaled). So, what is the reference group? The reference group is the response attribute that is assigned the value of 1. Selection of a reference group is based on how you would like to speak about your data. For example, suppose I am interested in women and attitudes towards gun violence, then the variable gender or sex, which has two response attributes would be coded as 1 = women and 0 = men. However, if I am interested in men and attitudes towards gun violence, then 1 = men and 0 = women. In other words, however you want to report your data, is how you decide on what the reference group is. The reference group establishes the point of analysis. It is very important to identify and label this reference group and even understand why it is important to the given analysis.

Correlations simply assess strength and direction of the relationship between variables, most often IVs and DV and is by far the most popular technique used to evaluate the association of two interval-ratio variables, most often the IV and DV. 'Statistically, it measures how much the scores of two variables vary' (Hinton, McMurray, Brownlow & Terry, 2024: 280). Sometimes, these are referred to as descriptive statistics as they simply report the trends and patterns of the associations amongst the IV and DV. A correlation is most often run prior to a regression analysis to explore the possible associations. Correlations have core functions which answer the following questions:

1   Is there a correlation or association between X and Y?
2   What is the direction of the relationship between X and Y? Positive or negative?
3   How strong is the relationship?
4   Is the relationship or correlation statistically significant?

One important substantial point is that correlation does not indicate causation. The statistical relationship or association between two key variables is quantified by changes in one variable to another. They simply determine association, either positive or negative by direction or strength via the correlation coefficient value or number (r). By examining the upper or lower portion of the identity matrix, the diagonal[1] allows us to assess each correlation coefficient with respect to the DV in the model. Only one side of the diagonal is reported as it is a mirror image of itself. Thus, select the upper or lower quadrant or half and begin to analyze the matrix. The most significant correlations to consider initially are the IVs or predictors associated with the DV. Correlation provides us with three key outcomes: strength of the relationship, directionality of the relationship and whether it is statistically significant or not. That is all. There is no cause and effect being established at this stage. To establish causation, which is rather complex, causal factors, logic and theory play a pivotal role. Additionally, temporal order of the variables, in which cause precedes effect, becomes critical, and one must always ensure that spurious effects through control variables are tested for. Without this, there is no causation established. For example, there may be a positive correlation between the GPA score and time spent studying. However, this correlation does not imply that GPA score causes people to study more or vice versa. Basically, an association has been formed and these two variables co-vary. This outcome provides a valuable insight about the data and the relationship but does not provide an established causal pathway. Careful consideration is required before establishing cause and effect. Sometimes, correlations are visually depicted. A *scatterplot* is the most appropriate graph for the visualization of the correlation coefficient.

The two basic elements of a correlation coefficient are strength and direction. First, let's review strength of a correlation. Strength in a correlation is determined by the numeric value of the correlation coefficient. It ranges from 0 to 1, 0 being no correlation and 1 being a perfect correlation. The closer the coefficient or r value is closer to 0 the weaker the relationship and the closer the coefficient or r value is to 1, the stronger the relationship. Correlations can be divided into four main groups: Perfect, Strong, Moderate and Weak and can be positive or negative. Please note that signage, positive or negative correlation, has nothing to do with strength. You can have a negative strong correlation or a positive weak correlation. Strength of a correlation has nothing to do with directionality of the correlation. A perfect correlation does not exist in the real world. Most often any variable correlated with itself yields a perfect correlation and formulates the diagonal in the correlation identity matrix. A strong correlation means that both variables are highly correlated, linear, and predictable. While the relationship is not perfect, there is an excellent straight-line relationship amongst the IV and DV. The predictive power is grand, and one variable can very effectively predict or explain a large amount of the variability in the other variable. A moderate correlation is suggestive of a mediocre predictable relationship in which there is evidence of a reasonable relationship that posits a middle ground. There is a moderate amount of predictability and linearity. It is not too weak nor too strong. Finally, a weak relationship is suggestive of a relationship that is close to 0, meaning no correlation. The IV and DV here are relatively negligible and not significantly correlated. Here, the relationship is non-linear and data points are randomly scattered with no consistent trend or pattern to report. There is chaos in the scatterplot and there is low predictive power in which on variable does not account for much of the variation in the other. A strong correlation most often is equivalent to less error and most data points fall close to the regression line or line of best bit, while a weak correlation is representative of more error and the points are increasingly scattered. The regression line or line of best fit represents the best estimation or approximation of the relationship between the IV and DV in a scatterplot. A strong correlation is always considered to be a better more 'fit' model for the data.

Most often the ranges of correlation values from perfect, strong, weak, to no correlation:

Size of Correlation or Correlation Coefficient (r) value	Interpretation
−1.00	Perfect negative correlation
−0.99 to −0.60	Strongly negative
−0.30 to −0.59	Moderately negative
−0.10 to −0.29	Weakly negative
0.00	No association or No correlation
+0.10 to +0.29	Weakly positive
+0.30 to +0.59	Moderately positive
+0.99 to +0.60	Strongly positive
+1.00	Perfect positive correlation

Another aspect of correlation is directionality. Direction of a correlation may either be positive or negative and is determined by signage (+/-). A positive correlation occurs when two variables increase or decrease simultaneously. For instance, as x increases, y increases; or as x decreases, y decreases. Examples of a positive correlation:

- As years of education increase, income increases
- As number of hours studying increases, GPA increases
- Gender increases number of children (Women are more likely to have more children)
- As anxiety scores increases, number of medications increase
- As number of convictions decrease, age of respondent decreases

A negative correlation is when two variables behave in the opposite direction. Thus, as x increases, y decreases; or as x decreases, y increases. For example:

- As years of education decreases, # of sexual partners decreases income increases
- Gender decreases # times victim of IPV (Men are less likely to be victims of IPV)
- As self-esteem decreases, weight increases
- As prison time increases, mental well-being decreases
- As years on the job increase, motivation decreases

Both positive and negative correlations operate uniquely, however both clearly show the nature of the association of variables and how a change in one variable affects the other. This is an important step in determining the type of association exists and the relationship amongst the variables, especially prior to running a regression analyses. While this information seems basic, it speaks volumes to the descriptive information it relays about an important relationship.

Sometimes in correlation testing, there is a vested interest in testing for 'control' variables, Z, with correlation coefficients. We are simply interested in if there is a spurious relationship amongst the original relationship. Partial and semi-partial correlations allow you to test for control variables to ensure that there are no third unseen variable at play. In partial correlations, the control variable impacts both X and Y or IV and DV. In semi-partial correlation, the control variable impacts only one of the variables, either X or Y or the IV or DV, not both. A similar 'r' or the correlation coefficient value among these variables simply means that the control variable plays no significant impact. However, a sizeable discrepancy in the 'r' value or correlation coefficients is indicative of the fact that the control variable is indeed influencing the relationship and thus, the control variable cannot be ignored, and the focus of the analysis becomes the control variable, rather than the relationship of the IV and DV. Adding a control variable to the correlation models allow for increased holistic explanations about the relationships and any confounding factor or effects. Both types of correlation become important in further assessing the nature of IVs with respect to the DV.

Furthermore, another important function of a Pearson Product moment correlation is that it tests for multicollinearity of two IVs prior to regression analyses. It is important that all IVs exhibit a *lack of multicollinearity*. This simply means that all IVs should bring their own unique variation to the model and there should be no sharing violation amongst IVs. Multicollinearity is a statistical concept in which two or more IVs or predictor variables in a model are highly correlated or associated, thus resulting in measuring similar concepts. This creates biased estimates which in turn creates a bias model. Any Pearson product moment correlation holding a value of greater than 0.80 is deemed multicollinear. Thus, one of the two predictor variables or IVs can be eliminated from the analyses such that multicollinear data is minimized. To run correlations, certain assumptions must be met.

Activity Alert

What does the Pearson Product Moment Correlation Measure?
Does it assess cause and effect?
What is a Partial Correlation vs. Semi-partial correlation?

Assumptions	Explanation
1. Data or level of measurement of variables	Correlation assumes that the independent and dependent variables are interval-ratio continuous or its equivalent (i.e., scaled or dummy coded; 1 is the reference group)
2. Sampling method and sample size	Random sampling and large sample
3. Independence of observations	Data observations must be independent of each other; no influence by other observations
4. Shape of the distribution	Normally distributed of bivariate relationship such that bivariate normality is met for each set of IV and DV
5. Homogeneity of variances	The variances are equal across all groups; spread of variability of scores or data is consistent for all IVs
6. Absence of multicollinearity	The IVs are not highly correlated with each other; each IV brings its own unique variance; each correlation is less than 0.80
7. Linearity	The relationship between the IVs and DV is linear
8. Confidence level	The amount of risk you are willing to take: 95%, 0.05 alpha
9. Sample statistic	'r' correlation coefficient; rejects the H_o or fails to reject it to see overall significance of the IVs alongside DV
10. Outliers	Minimize outliers or extreme scores

What variables are required to run this test?

- At least one IV or predictor: interval-ratio continuous or scaled or dummy coded variables, assigning values of 0 and 1
- Dependent Variable: 1 Interval-ratio continuous or scaled.

Null and Research Hypotheses

Null hypothesis: There are no statistically significant correlations or associations amongst the IV(s) and DV.

Research hypothesis: There are statistically significant correlations or associations amongst the IV(s) and DV.

The sample statistic here is the Pearson product moment correlation (r). The power of the Pearson coefficient is to determine the extent of the strength, direction, and significance of the association for the continuous variables in question, most often the IV(s) and DV. These associations determine and describe the nature of the plausible causal linkages or associations prior to running a regression model. This statistic, although highly descriptive in nature, becomes quite important in establishing the potential associations and/or correlations. It provides a great space and map of what one should expect in a regression model. Depending on the various outcomes, certain informed decisions can be made at this stage in establishing the final causal model that you may want to eventually test via regression analysis. At this stage, you can also explore instances of multicollinearity amongst various IVs and ensure two IVs are not too heavily correlated and each IV brings its own unique variation to the model. Thus, some model specification and re-specification can take place. The sample statistic here is correlation coefficient, r.

$$r = \frac{\sum (x_i - \bar{x})(y_i - \bar{y})}{\sqrt{\sum (x_i - \bar{x})^2 \sum (y_i - \bar{y})^2}}$$

$$r = correlation\ coefficient$$
$$x_i = values\ of\ the\ x-variable\ in\ a\ sample$$
$$\bar{x} = mean\ of\ the\ values\ of\ the\ x-variable$$
$$\bar{y}_i = values\ of\ the\ y-variable\ in\ a\ sample$$

Key Statistics to Report

1. Report each Correlation coefficient with strength, direction, and significance on either side of the diagonal of the identity matrix as the IVs relate to the DV
2. Report any issues with Multicollinearity amongst the IVs. Any coefficient greater than 0.80 should be removed or reconsidered

Pearson Product Moment Correlation in SPSS

1   Click on Analyze > Correlate > Bivariate
2   Move the ALL variables of interest into the Variables Box
3   Ensure Pearson is checked off and Two-tailed
4   OK to generate a Pearson Product Moment Correlation

Partial Correlation in SPSS (with Control)

1   Click on Analyze > Correlate > Partial
2   Move the ALL variables of interest into the Variables Box
3   Add CONTROL variable >Controlling for: box
4   In Options > Check off: Zero-order Correlations
5   OK to generate a Partial Correlation

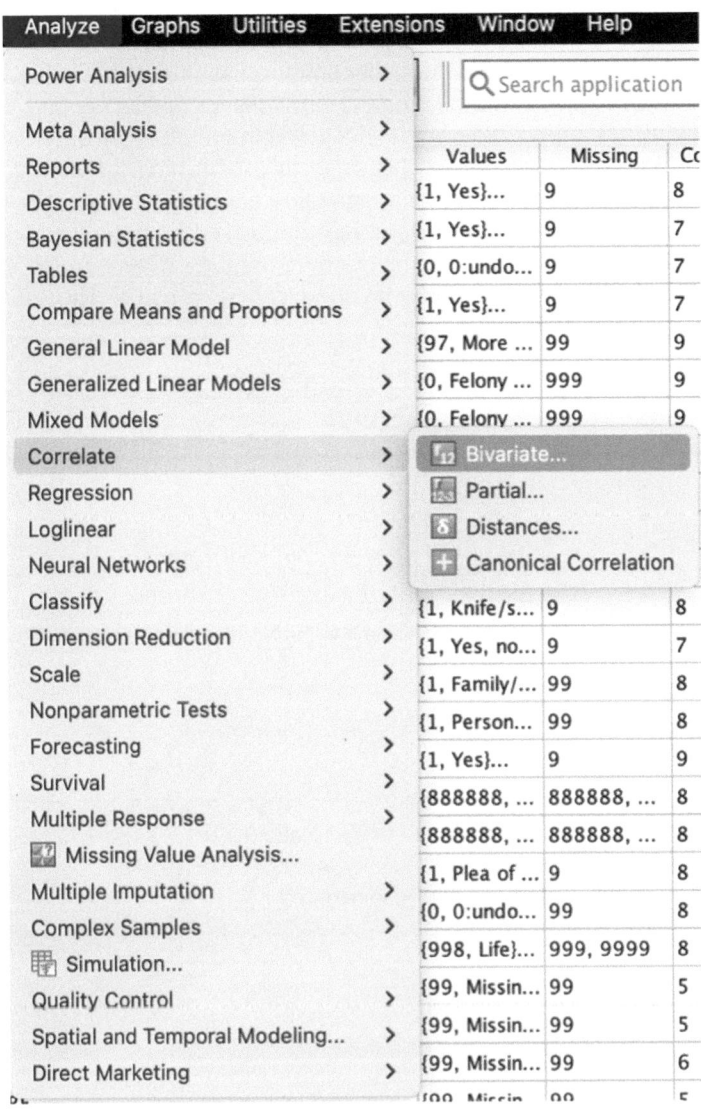

Figure 10.1a SPSS Command for Pearson Product Moment Correlation

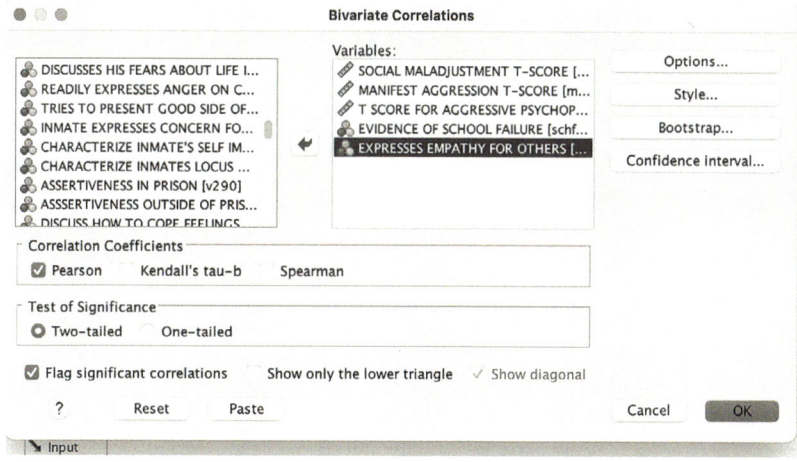

Figure 10.1b Move the ALL variables of interest into the Variables Box and Ensure Pearson is checked off and Two-tailed

Descriptive Statistics

	Mean	Std. Deviation	N
SOCIAL MALADJUSTMENT T–SCORE	60.47	16.075	228
MANIFEST AGGRESSION T–SCORE	46.69	10.467	228
DCschfail2 (1=no)	.54	.500	228
empathy 2 DC (1=empathetic)	.81	.392	228
T SCORE FOR AGGRESSIVE PSYCHOPATH SCALE	47.58	6.784	228

Correlations

		SOCIAL MALADJUSTMENT T–SCORE	MANIFEST AGGRESSION T–SCORE	DCschfail2 (1=no)	empathy 2 DC (1=empathetic)	T SCORE FOR AGGRESSIVE PSYCHOPATH SCALE
Pearson Correlation	SOCIAL MALADJUSTMENT T–SCORE	1.000	.756	−.302	−.203	.421
	MANIFEST AGGRESSION T–SCORE	.756	1.000	−.211	−.164	.381
	DCschfail2 (1=no)	−.302	−.211	1.000	.139	−.439
	empathy 2 DC (1=empathetic)	−.203	−.164	.139	1.000	−.228
	T SCORE FOR AGGRESSIVE PSYCHOPATH SCALE	.421	.381	−.439	−.228	1.000
Sig. (1–tailed)	SOCIAL MALADJUSTMENT T–SCORE	.	<.001	<.001	.001	<.001
	MANIFEST AGGRESSION T–SCORE	.000	.	.001	.007	.000
	DCschfail2 (1=no)	.000	.001	.	.018	.000
	empathy 2 DC (1=empathetic)	.001	.007	.018	.	.000
	T SCORE FOR AGGRESSIVE PSYCHOPATH SCALE	.000	.000	.000	.000	.
N	SOCIAL MALADJUSTMENT T–SCORE	228	228	228	228	228
	MANIFEST AGGRESSION T–SCORE	228	228	228	228	228
	DCschfail2 (1=no)	228	228	228	228	228
	empathy 2 DC (1=empathetic)	228	228	228	228	228
	T SCORE FOR AGGRESSIVE PSYCHOPATH SCALE	228	228	228	228	228

Figure SPSS OUTPUT #10 Pearson Product Moment Correlation

Technical and Substantive Interpretation

Measure of Association: Pearson Product Moment Correlation

Research Question: What are the associations or correlations amongst the IVs or predictor variables and the dependent variable? For instance, are there significant, positive or negative associations amongst the DV, social maladjustment score, with manifest aggression score, school failure, empathy, and aggressive psychopath?

Null Hypothesis: There are no statistically significant associations or correlations amongst the IVs and/or predictors with the DV, social maladjustment score.

Research Hypothesis: There are statistically significant associations or correlations amongst the IVs and/or predictors with the DV, social maladjustment score.

Technical Interpretation

Descriptive statistics: The sample size for the correlation was reasonable at about 228 respondents. The means for the interval-ratio variables indicate that on average social maladjustment, manifest aggression and aggressive psychopathy scales had scores of 60.47, 46.69, and 47.58 respectively.

Pearson Product Moment Correlations (r): The correlation coefficient for the DV, social maladjustment, and the IV manifest aggression was 0.756, < 0.001 ($p < 0.05$). There was a positive strong significant correlation between social maladjustment and manifest aggression. As social maladjustment increased, manifest aggression increased as well. The correlation coefficient between social maladjustment and school failure was -0.302, < 0.001 ($p < 0.05$). There was a negative moderate significant correlation amongst the variables. Those who did not fail school scored lower on the social maladjustment score. Thus, they were less likely to violate social norms. The correlation coefficient between social maladjustment and empathy had a value of -0.203, < 0.001 ($p < 0.05$) indicating a negatively weak yet significant correlation. Empathetic respondents received a lower score on the social maladjustment score. Therefore, empathetic beings were less likely to violate social norms. Finally, the correlation with social maladjustment and aggressive psychopath was 0.421, < 0.001 ($p < 0.05$). This was a positive moderate yet significant association. Accordingly, as social maladjustment increased, aggressive psychopathic tendencies also increased. There were no instances of multicollinearity. No correlation amongst the IVs was greater than 0.80. Thus, no sharing violations and each predictor was bringing its own unique variation to the model.

Substantive Interpretation

Overall, the results of this correlation or identity matrix were expected. The correlations clearly demonstrated an expected finding. Social maladjustment was highly correlated with manifest aggression, followed by aggressive psychopath, school failure and empathetic tendencies. Aggressive respondents scored higher on the social maladjustment variable, compared to those who did not fail school and were empathetic. These results overall were expected to a certain degree. The results of the correlation provide great insight into how the regression model will play out and what we can expect in future analyses.

Final Thoughts

Correlation, a measure of association for interval-ratio continuous data is a great statistical tool for describing relationships. The value correlations bring to data analysis is vast. Without correlations there is no way to figure out how two variables co-vary or how they are correlated. Knowing how two variables correlate, allows to make better informed decisions about the more advanced statistical techniques and the statistical pathways that can be taken. Correlations allow us to see variables from a different perspective. Knowing strength, direction, and significance of a relationship speaks volumes to the variables in questions. Specific causal modelling can be done based on the correlation coefficient values. Correlations allow us to confirm observations and specify and re-specify the models being tested for regression analyses.

No statistical technique is flawless. At this point, some of the limitations of correlation are worth mentioning, despite what it offers to statistics. One of the major limitations is that correlation does not establish cause and effect or imply it. Just because two variables are correlated does not mean that one variable causes the other. More complex analyses, like regression, are prudent to establish cause and effect and require controlling for variables. Without a linear relationship, correlation may not accurately capture the association. In such cases, alternative statistical methods may be needed to analyze the data. Correlations, without testing for controls, is limited to bivariate relationships. The social world comprises of many complex relationships and/or variables and thus relying solely on correlations for a holistic analysis may not be fruitful. Advanced statistical techniques may be necessary to get at the complexities of the relationships. Additionally, correlations are sensitive to outliers or extreme values which may skew and biased results. Normal distributions are critical to correlation analysis as it is a parametric test. If the variables do not conform to normal distributions, an alternative such as Spearman's rank correlation may be more appropriate. This non-parametric correlation method does not rely on the data adhering to assumptions of normality and linearity. Another limitation of correlations is that for nominal variables, it necessitates recoding into binary dummy codes (0 and 1), which can be a labor-intensive task. Furthermore, correlations require a substantial sample size is essential to generate reliable and effective correlation coefficients. While correlation analysis is a valuable tool for exploring relationships among interval-ratio continuous variables, it is not without limitations. Researchers should exercise caution to avoid conflating correlation with causation, thoroughly scrutinize underlying assumptions, and consider alternative methods when dealing with complex relationships or datasets with unique characteristics. Overall, correlations are a useful tool for exploring interval-ratio continuous variable relationships but does come with its limitations. Researchers should be cautious about inferring causation, assess assumptions, and consider alternative methods when dealing with complex relationships or datasets with specific characteristics. At the end correlation analysis is a basic tool to address simple bivariate associations and understand the nuances of a relationship between the IV and DV. At the heart of correlation lies the strength and direction of any relationship. While correlation provides information and important information, it is not so statistically powerful in making final remarks on causal modelling it can only guide those models to a certain degree. The final chapter in this book reviews causal model regression analysis which is the ultimate statistical design test that better understands causality and causal linkages in relationships.

Keywords and Definitions

Association	This refers to the simple correlation of two variables or the association amongst them, not necessarily the cause and effect or causation.

The Pearson Product Moment Correlation (r)	A bivariate measure of association that is heavily linked to interval-ratio continuous data to assess strength and direction of the relationship between the IV and DV or X and Y. Correlation does not indicate causation.
Strength	The strength in a correlation is determined by the numeric value of the correlation coefficient. It ranges from 0 to 1, 0 being no correlation and 1 being a perfect correlation. The closer the coefficient or r value is to 0, the weaker the relationship and the closer the coefficient or r value is to 1, the stronger the relationship.
Direction	The direction of a correlation may either be positive or negative and is determined by signage (+/–). A positive correlation occurs when two variables increase or decrease simultaneously. For instance, as x increases, y increases; or as x decreases, y decreases.
Multicollinearity	A statistical concept in which two or more IVs or predictor variables in a model are highly correlated or associated, thus resulting in measuring similar concepts and having a sharing violation. If the coefficient is 0.80 or higher, multicollinearity has taken place and either variable should be removed from the analysis.
Partial correlations	These are correlations in which the control variable impacts both X and Y or IV and DV.
Semi-partial correlation	These are correlations in which the control variable impacts either X or Y or IV or DV.

Test Your Knowledge

1   ____ is a correlation that assesses strength and direction of the relationship for interval-ratio continuous data amongst the IV and DV.

   a   Partial correlation
   b   Semi-partial correlation
   c   Pearson Product Moment Correlation
   d   Regression
   e   a, b, c

2   The ___ of the relationship is determined by the r value and numerical coefficient.

   a   Direction
   b   Strength
   c   Both direction and strength
   d   Prediction
   e   Association

3   The ___ of the relationship is determined by the relationship being positive or negative.

   a   Strength
   b   Direction
   c   Variation

d   Prediction
e   Association

4 As the # of hours sleep increase, there is a decrease in overall productivity is an example of a _____ relationship.

a   Positive
b   Negative
c   Perfect
d   a and b
e   None of the above

5 As GPA increases, there is increased academic hope. This is an example of a _____ correlation.

a   Positive
b   Negative
c   Perfect
d   a, b, c
e   None of the above

6 While correlations test for strength and direction of an IV and DV, they also test for any shared violations amongst the IVs. When two IVs are highly correlated (r is > 0.80) that is referred to as:

a   Homoscedasticity
b   Multicollinearity
c   Regression
d   Partial correlation
e   Assumption

7 A _____ correlation tests for a control variable that impacts the IV and DV both.

a   Semi-partial
b   Pearson Product Moment Correlation
c   Partial correlation
d   Regression
e   Multicollinear

8 A ___ correlation tests for a control variable that impacts either the IV or DV.

a   Partial correlation
b   Pearson Product Moment Correlation
c   Semi-partial correlation
d   Regression
e   Multicollinear

9 The correlation family is considered a descriptive statistic that is a bivariate measure of association.

a   True
b   False

10 Correlation are statistically significant, meaning they significantly co-vary and two variables are indeed related when:

a $p < 0.05$
b $p > 0.05$
c $p = 0.05$
d   Correlations can never be significant
e   Correlations are linear

Note

1 The diagonal represents each variable correlated with itself and thus assumes a value of 1.00 or perfect correlation. Any variable correlated with itself assumes a perfect correlation of 1.00. Use this diagonal to divide the upper and lower half of the correlation table and begin interpretations of your IVs and DV.

11 A World of Prediction(s) Using Regression Analyses
Multiple Linear Regression, Hierarchical Regression, and Logistic Regression

We begin this final chapter by looking into the world of predictions using data. The focus of this final chapter is the statistical analysis by means of variety of regression models. This book culminates with an examination of Bivariate regression, Multivariate or Multiple linear regression, Hierarchical or Incremental or Block modelling regression and finally, Logistic regression. In examining the social world around, us we may want model relationships with specific variables and see how an IV(s) relates to a specific DV; or we may want to make a prediction; or we may want to test a specific hypothesis; or be curious about certain control variables and establish a true causal linkage. All the regression analyses mentioned in this chapter do just this in their own special way and their outcomes may inform certain policy decisions, quality control, risk assessment, academic performance, marketing, etc. A whirlwind of possibilities is possible with these tests. These analyses come full circle in statistical or data analysis and really complete the entire storytelling process. Soon after the Pearson Product Moment correlations are run, assessed, and evaluated, various regression analyses can be run. Why you choose a specific statistical test depends heavily on the nature of your variables and characteristics of the distributions.

The Liner regression model is the test that becomes the natural and accepted passage to these statistical analyses. There are many kinds of regressions, like Logistic regression and Hierarchical or Incremental or Block modelling[1] regression models but the Multiple linear regression model is by far the most popular and widely used by social scientists. The main goal of these tests is twofold: predictability and explanation. This statistical test is much used in the social sciences to assess and establish causal linkages.

The two *most* common regression models in the humanities and social sciences: Bivariate linear regression and Multivariate or Multiple linear regression. These two statistical analyses are the statistical footprint of interval-ratio continuous data. Linear bivariate or multivariate or multiple regression, also known as ordinary least squares (OLS) method, is one of the most sophisticated inference-based statistical tests which is used with the highest level of measurement, interval-ratio continuous data or its equivalent (i.e., dummy coded or scaled variables) to assess causal linkages. Regression analysis requires both IV and DV to assume interval-ratio characteristics and works with the highest level of measurement. This type of regression analysis may be bivariate, have only one IV or predictor variable or involve multiple IVs or predictors, hence, the name multiple linear regression analysis. IVs are often referred to as predictor variables and the DV as the criterion or outcome. Regression goes one step further then correlation, in that it specifies the exact nature of the relationship in question by allowing researchers to make predictions but also assesses variation and explanatory power of the model. In regression, the onus is on the researcher to

DOI: 10.4324/9781003215691-14

adequately specify the model. This entails including relevant predictors in the model. Any predictor variables included in the model are based on a valid theoretical social explanation, rather than empirical evidence alone. The model is specified and respecified accordingly.

The purpose of this test is two-fold. First, it is the statistical test that makes estimated predictions. Second, it tells us the exact nature of the variables and further assesses the explanatory power of the model. By assessing the overall fit or explained variance of the model (R^2 or Adjusted R^2) out of 100%, great information is revealed about the analysis. In regression, you achieve overall model contribution or relative model contribution. The four essential functions of regression analysis are the following:

1   It allows us to predict an individual's score on one variable based on knowledge of other variables using $y = a + \beta x$ (bivariate model) or $y = a + \beta_1 x_1 + \beta_2 x_2 \beta_3 x_3 + \beta_n x_n$, where a is the constant and β is the unstandardized β value
2   It allows us to assess whether certain predictors or IVs can explain a particular outcome well via the regression coefficients (i.e., unstandardized βs and standardized beta values)
3   It assesses the explained variance out of 100% and assesses model fit
4   It only assesses direct effects

Most regression models, except Logistic regression,[2] are based on several assumptions that must be met for the accuracy of the model. Some noteworthy assumptions and/or requirements prior to running a regression to pay attention to are: So, what is the difference? In a true bivariate model, there are a total of two variables, one X and one Y that represent the IV or predictors and DV or outcome respectively. The multiple regression model is more complex modelling and relationship building and testing in which there are multiple X's or IVs and one DV.

Characteristics	Bivariate Regression Modelling	Multiple Linear Regression Modelling
1. Number of variables	Only two variables; one IV and one DV	Multiple IVs and one DV
2. Relationship type	Simple linear relationship	Complex linear relationships
3. Prediction equation	$\hat{y} = a + \beta x + error$	$\hat{y} = a + a + \beta_1 x_1 + \beta_2 x_2 \beta_3 x_3 + \beta_n x_n + error$
4. Interpretation	Simple	Complex

Both these regression analyses, although similar, have unique characteristics. This is the ultimate inference-based test as it provides us with so much information about the models we are testing. This statistical technique and tool allow one to see the relative influence of one variable on another. For example, aggression scale score, empathetic expression, family network and racial background may influence the number of times you end up in jail. These types of analyses take on multiple predictors to explain a particular social outcome. However, they like other inference-based tests are highly assumption bound. These assumptions must be met for regression analyses to be reliable and valid, as well as to enhance generalizability.

Activity Alert

What is the difference between bivariate and multivariate linear regression?

Model Specification, Goodness of Fit vs. Model Fit

Regression entails model specification and fully understanding how the direct effects of the predictor variable(s) and DV operate in the model. Your models should be theoretically sound and conceptualized accordingly. They should make logical sense to make the best fitting regression model. Additionally, this type of statistical analysis evaluates 'goodness of it' and 'model fit'. Essentially regression analysis or models carefully assess the underlying relationships between the IVs or predictors and the DV or outcome being measured. The goodness of fit evaluates how well the regression model's predictions match the actual data or the distance between the observed and predicted values. The Coefficient of determination or Adjusted R^2 assesses model fit by seeing how much variation in the DV is explained by the predictor(s) or IV(s). Examining the error or difference between observed and predicted values also provide trends and patterns that perhaps, indicate a lack of fit. Sometimes, this is called a residual analysis. Therefore, the relationship between the observed and predicted values becomes relevant and critical to understanding the workings of the overall model. Model fit, on the other hand conveys how well the selected predictors work for the model. Basically, do we have decent predictors in the regression model? Is our selection of IV(s) or predictors accurate and do they make a meaningful contribution to the overall model? The Global F-test in regression, also known as the ANOVA F-test, if statistically significant, indicates that the selected predictor(s) for the model are decent for that DV indicating the overall model fit is good and meaningful. A non-significant outcome indicates a misfit of variables and thus, the model needs to be reconceptualized or modified accordingly. Moreover, ensuring no assumptions are violated for the regression analysis also enhances overall model fit. There are a handful of assumptions that must not be violated for the regression to be reliable, valid, and generalizable from a sample to a population. Overall, bivariate, or multi-variate linear regression models really emphasize predictive power, a model that accurately predicts the DV, and meeting all or most assumptions to maximize good-ness of it and model fit. Let's see what the assumptions are in the next section.

Assumptions	Explanation
1. Data or level of measurement of variables	Regression analysis assumes that the independent and dependent variables are interval-ratio continuous or its equivalent (i.e., scaled, or dummy coded; 1 is the reference group)
2. Sampling method and sample size	Random sampling and large sample
3. Independence of observations	Data observations must be independent of each other; no influence by other observations
4. Shape of the distribution	Normally distributed of bivariate relationship such that bivariate normality is met for each set of IV and DV
5. Homoscedasticity	Constant spread of variance (equal variances for residuals or error terms)

(continued)

Assumptions	Explanation
6. Absence of multicollinearity	The IVs are not highly correlated with each other; each IV brings its own unique variance; Each correlation is less than 0.80. Variance inflation factor > 2.50; or Tolerance values < 0.40
7. Linearity	The relationship between the IVs and DV is linear
8. Confidence level	The amount of risk you are willing to take: 95%, 0.05 alpha
9. Sample statistic	t-statistic or t-value; rejects the H_o or fails to reject it to see overall significance of the IVs alongside DV
10. Outliers	Minimize outliers or extreme scores using Casewise Diagnostics, Mahalanobis Distance and Cooks Distance

These assumptions are critical to goodness of fit and model fit of data.[3]
What variables are required to run this test?

- At least one or multiple IVs or predictors: interval-ratio continuous or scaled or dummy coded variables, assigning values of 0 and 1, with 1 being the reference group
- Dependent variable: One interval-ratio continuous or scaled.

The research question and hypotheses are written in terms of predictability and whether a certain IV or predictor significantly predicts an outcome or DV. Each bivariate and multivariate model has its own arrow diagram or causal model.

Bivariate Model

Research Question

Is there a statistically significant predictor for the DV?
   Years of Education predicts Current Salary

Null and Research Hypotheses

Null hypothesis: There is no statistically significant predictor or IV for the DV. Years of Education is not a statistically significant predictor of Current Salary.

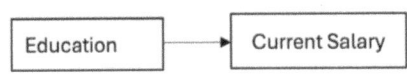

Figure 11.1 Arrow Diagram or Causal Model for a Bivariate Model

Research hypothesis: There is statistically significant predictors or IV for the DV. Years Education is a statistically significant predictor of Current Salary.

Regression equation or Line of best fit for making prediction: $\hat{y} = a + \beta x + error$

Equation terms	
$\hat{y}$	$\hat{y}$ is the 'predicted' outcome [our dependent variable]
a	a is the y-intercept or constant (point where regression line intersects the ordinate x=0)
Unstandardized b	β represents the slope of the regression line (the amount of change in y that is associated with a unit change in x)
x	x is the score on the predictor variable(s) or IV(s)
e	e is the error term (residual)

Using the values from the regression coefficients table in SPSS to solve the prediction.

Multivariate Model

Research Question: Are there statistically significant predictors for the DV?
Years of Education, job category, and gender predict Current Salary
Null and Research hypotheses
Null hypothesis: There are no statistically significant predictors or IVs for the DV. Years of Education, job category, and gender are not statistically significant predictors of Current Salary.

Research hypothesis: There are statistically significant predictors or IVs for the DV. Years Education, job category, and gender are not statistically significant predictors of Current Salary.

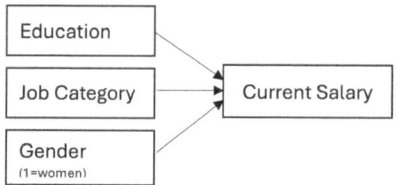

Figure 11.2 Arrow Diagram or Causal Model for a Multivariate Model

Regression equation or Line of best fit for making prediction:
$\hat{y} = a + \beta_1 x_1 + \beta_2 x_2 \beta_3 x_3 + \beta_n x_n + error$

Equation terms	
$\hat{y}$	$\hat{y}$ is the 'predicted' outcome [our dependent variable]
a	a is the y-intercept or constant (point where regression line intersects the ordinate x=0)
Unstandardized β	$\beta_1 x_1 + \beta_2 x_2 \beta_3 x_3 \ldots + \beta_n x_n$ represents the slope of the regression line (the amount of change in y that is associated with a unit change in x)
x	$x_1, x_2, x_3 \ldots x_n$ is the score on the predictor variable(s) or IV(s)
e	e is the error term (residual)

Using the values from the regression coefficients table in SPSS to solve the prediction.
The sample statistic, the t-value, determines whether there are significant predictors in the model and if the null hypothesis is fully rejected or partially rejected using a 95% confidence level and an alpha of 0.05.

Key Statistics to Report

1   Descriptive statistics, sample size, averages, means, and standard deviation are reported so that trends and patterns of the variables can be understood; you would assess all interval-ratio or scaled variables

2   Adjusted R^2 (explained variance) or the coefficient of determination explains how much variance in the DV is being explained by certain predictors out of 100%

3   Global F-test (decent predictors in the model) and must be statistically significant; this indicates that there are decent predictors in the model or the way the model has been conceptualized is correct; if this value is not significant then the model needs to be reconceptualized

4   Regression coefficients

   a   Unstandardized β (beta) tells the story of each predictor with respect to the DV; it tells us for every unit increase/decrease in x, y increases by a certain amount; if the value is positive, it is an increase; if the value is negative it is decrease; the unstandardized β is in units of the DV always

   b   Standardized beta, often denoted as β (beta), taking the absolute value (regardless of signage), tells the ORDER of best predictors as it relates to the DV. They are on the same scale and thus can be compared. Unstandardized β's cannot do this

   c   Sample statistic: t-value is the sample statistic that is the measure that assesses statistical significance of each regression coefficient as it relates to the DV. This statistic tests the null hypothesis and rejects or fails to reject the null hypothesis. Using a 95% confidence level, if the $p < 0.05$ then statistical significance is achieved and the null hypothesis is rejected; if the $p > 0.05$, then statistical significance is not achieved, and one fails to reject the null hypothesis

   d   Collinearity diagnostics: assess multicollinearity among IVs (predictors) in a regression model. Multicollinearity occurs when two or more predictors have a sharing violation, and no unique variance is present. The correlation amongst predictors is strong to include in the model. Correlation (> 0.80), Variance inflation factor (> 2.50) and Tolerance (< 0.40) indicate multicollinearity.

5   Outliers: Casewise Diagnostics, Mahalanobis Distance (Mahal's D) and Cook's Distance. Extreme scores, high or low, in the data; Mahal's Distance critical value must not be exceeded; Cook's Distance any value greater than 1.00 indicates outliers make an undue influence

6   Review Histogram and Normal p-p plot for examining normality and straight-line relationships for the predictors and DV and residuals

Multiple Linear Regression in SPSS

1   Click on Analyze >Regression > Linear

2   Move the variables of interest into the Dependent (DV) and Independent Variable Boxes Simultaneously

3   Click on Statistics >Check off Estimates, Model Fit, Descriptives, and Collinearity Diagnostics; Check-off Casewise Diagnostics and Click on Continue

4   Click on Plots >Y: Add ZPRED and X: Add ZRESID->Next check of Histogram and Normal Probability Plot >Click on Continue

5   Click on Save >Check off Mahalanobis Distance and Cook's Distance

6   OK to generate a Multiple Linear Regression

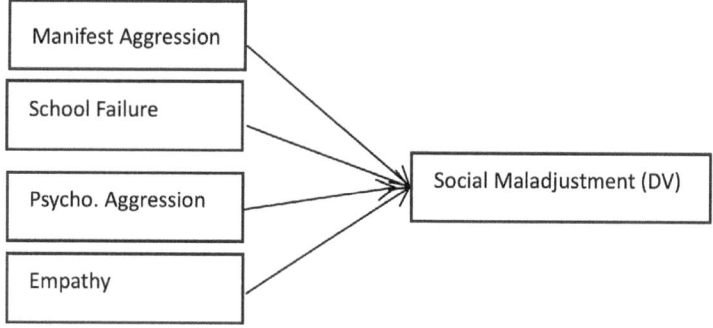

Figure 11.3a Multiple Linear Regression Causal Model of Manifest Aggression, School Failure, Psychopathic Aggression, and Empathy as Predictors of Social Maladjustment

Analyze	Graphs	Utilities	Extensions	Window	Help			
Power Analysis	>		Q Search application					
Meta Analysis	>							
Reports	>		Values	Missing	Columns	A	
Descriptive Statistics	>		{1, Yes}...	9	8	≣ Ri	
Bayesian Statistics	>		{1, Yes}...	9	7	≣ Ri	
Tables	>		{0, 0:undo...	9	7	≣ Ri	
Compare Means and Proportions	>		{1, Yes}...	9	7	≣ Ri	
General Linear Model	>		{97, More ...	99	9	≣ Ri	
Generalized Linear Models	>		{0, Felony ...	999	9	≣ Ri	
Mixed Models	>		{0, Felony ...	999	9	≣ Ri	
Correlate	>		{0, Felony ...	999	9	≣ Ri	
Regression	>		Automatic Linear Modeling...			;i	
Loglinear	>		Linear...			;i	
Neural Networks	>		Linear OLS Alternatives	>		;i	
Classify	>		Curve Estimation...			;i	
Dimension Reduction	>		Partial Least Squares...			;i	
Scale	>						
Nonparametric Tests	>		Binary Logistic...			;i	
Forecasting	>		Multinomial Logistic...			;i	
Survival	>		Ordinal...			;i	
Multiple Response	>		Probit...			;i	
Missing Value Analysis...							
Multiple Imputation	>		Nonlinear...			;i	
Complex Samples	>		Weight Estimation...			;i	
Simulation...			2-Stage Least Squares...			;i	
Quality Control	>						
Spatial and Temporal Modeling...	>		Quantile...			;i	
Direct Marketing	>		Optimal Scaling (CATREG)...			;i	
)RE			Kernel Ridge...			;i	
-SCORE			{99, Missin...	99	5	≣ Ri	

Figure 11.3b SPSS Command for Linear Multiple Regression

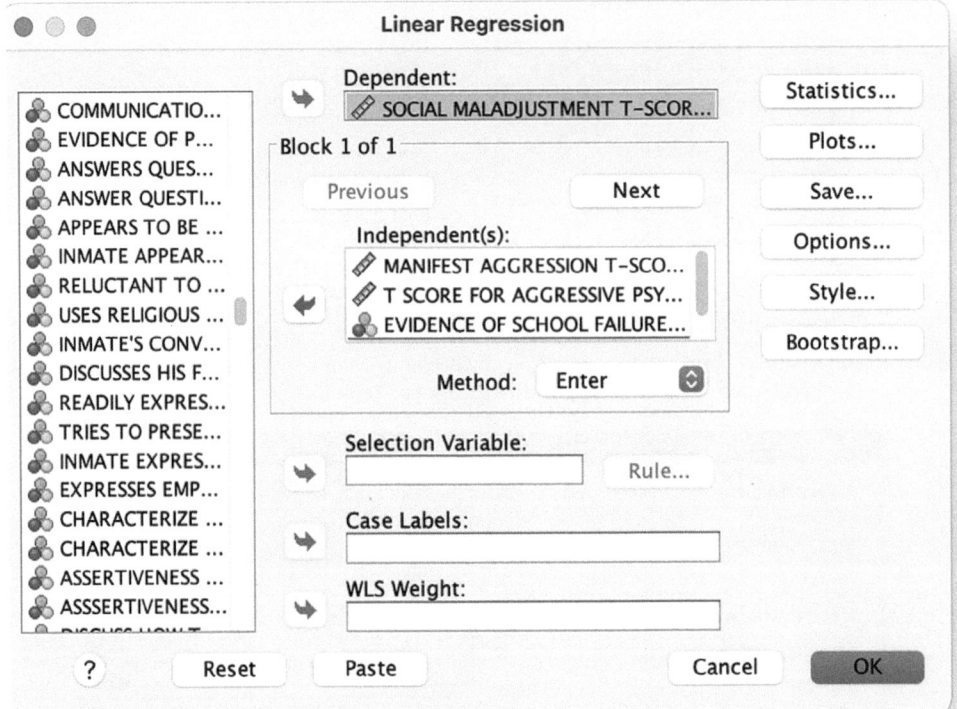

Figure 11.3c Move the variables of interest into the Dependent (DV) and Independent Variable Boxes Simultaneously

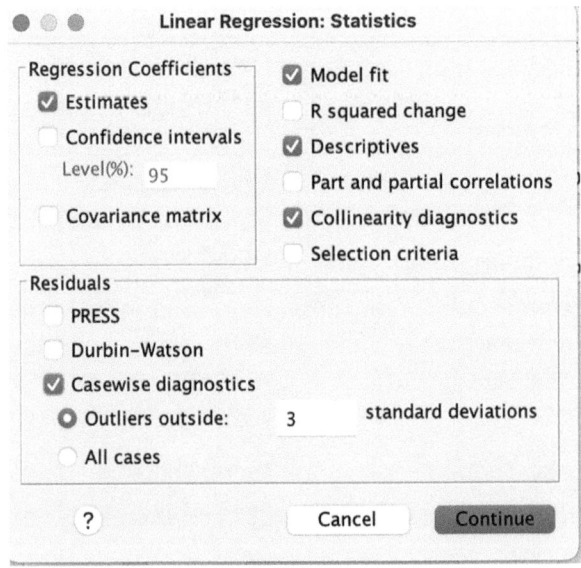

Figure 11.3d Click on Statistics, Check-off Estimates, Model Fit, Descriptives, and Collinearity Diagnostics; Check-off Casewise Diagnostics and Click on Continue

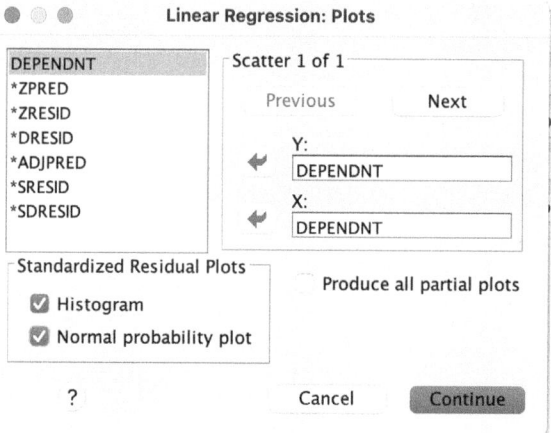

Figure 11.3e Click on Plots, Y: Add ZPRED and X: Add ZRESID and check off Histogram and Normal Probability Plot

Figure 11.3f Click on Save, Check-off Mahalanobis Distance and Cook's Distance to get Outlier Assessment

Descriptive Statistics

	Mean	Std. Deviation	N
SOCIAL MALADJUSTMENT T–SCORE	60.47	16.075	228
MANIFEST AGGRESSION T–SCORE	46.69	10.467	228
DCschfail2 (1=no)	.54	.500	228
empathy 2 DC (1=empathetic)	.81	.392	228
T SCORE FOR AGGRESSIVE PSYCHOPATH SCALE	47.58	6.784	228

Figure SPSS Output #11.1 Linear Multiple Regression Model

Variables Entered/Removed[a]

Model	Variables Entered	Variables Removed	Method
1	T SCORE FOR AGGRESSIVE PSYCHOPATH SCALE, empathy 2 DC (1=empatheti c), MANIFEST AGGRESSION T–SCORE, DCschfail2 (1=no)[b]	.	Enter

a. Dependent Variable: SOCIAL MALADJUSTMENT T–SCORE

b. All requested variables entered.

Model Summary[b]

Model	R	R Square	Adjusted R Square	Std. Error of the Estimate
1	.777[a]	.604	.597	10.211

a. Predictors: (Constant), T SCORE FOR AGGRESSIVE PSYCHOPATH SCALE, empathy 2 DC (1=empathetic), MANIFEST AGGRESSION T–SCORE, DCschfail2 (1=no)

b. Dependent Variable: SOCIAL MALADJUSTMENT T–SCORE

ANOVA[a]

Model		Sum of Squares	df	Mean Square	F	Sig.
1	Regression	35404.928	4	8851.232	84.896	<.001[b]
	Residual	23249.857	223	104.259		
	Total	58654.785	227			

a. Dependent Variable: SOCIAL MALADJUSTMENT T–SCORE

b. Predictors: (Constant), T SCORE FOR AGGRESSIVE PSYCHOPATH SCALE, empathy 2 DC (1=empathetic), MANIFEST AGGRESSION T–SCORE, DCschfail2 (1=no)

Coefficients[a]

Model		Unstandardized Coefficients B	Unstandardized Coefficients Std. Error	Standardized Coefficients Beta	t	Sig.	Collinearity Statistics Tolerance	Collinearity Statistics VIF
1	(Constant)	3.427	6.448		.531	.596		
	MANIFEST AGGRESSION T–SCORE	1.053	.070	.686	14.968	<.001	.847	1.181
	DCschfail2 (1=no)	-3.386	1.513	-.105	-2.237	.026	.804	1.244
	empathy 2 DC (1=empathetic)	-2.181	1.784	-.053	-1.223	.223	.939	1.064
	T SCORE FOR AGGRESSIVE PSYCHOPATH SCALE	.241	.119	.102	2.022	.044	.704	1.421

a. Dependent Variable: SOCIAL MALADJUSTMENT T–SCORE

Casewise Diagnostics[a]

Case Number	Std. Residual	SOCIAL MALADJUSTMENT T-SCORE	Predicted Value	Residual
19	3.179	81	48.54	32.455
60	3.117	81	49.18	31.824
118	3.344	81	46.86	34.141
278	4.160	81	38.52	42.478

a. Dependent Variable: SOCIAL MALADJUSTMENT T–SCORE

Residuals Statistics[a]

	Minimum	Maximum	Mean	Std. Deviation	N
Predicted Value	38.52	97.44	60.47	12.489	228
Std. Predicted Value	-1.757	2.960	.000	1.000	228
Standard Error of Predicted Value	.977	3.029	1.468	.363	228
Adjusted Predicted Value	37.72	98.82	60.47	12.507	228
Residual	-30.548	42.478	.000	10.120	228
Std. Residual	-2.992	4.160	.000	.991	228
Stud. Residual	-3.022	4.199	.000	1.003	228
Deleted Residual	-31.163	43.281	.002	10.357	228
Stud. Deleted Residual	-3.079	4.366	.002	1.013	228
Mahal. Distance	1.084	18.981	3.982	2.644	228
Cook's Distance	.000	.091	.005	.011	228
Centered Leverage Value	.005	.084	.018	.012	228

a. Dependent Variable: SOCIAL MALADJUSTMENT T–SCORE

Charts

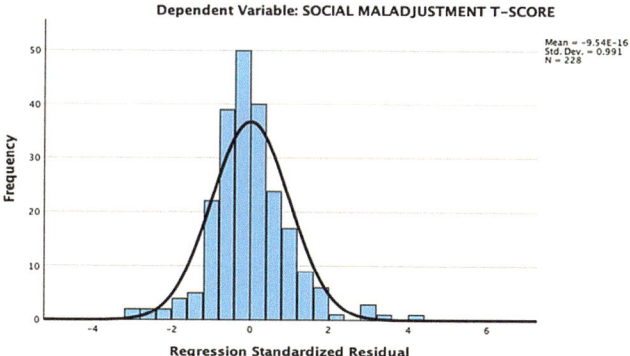

Histogram
Dependent Variable: SOCIAL MALADJUSTMENT T–SCORE

Mean = -9.54E-16
Std. Dev. = 0.991
N = 228

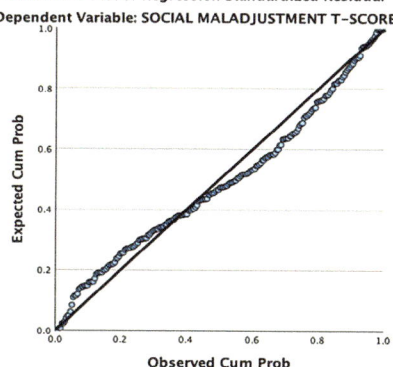

Normal P–P Plot of Regression Standardized Residual
Dependent Variable: SOCIAL MALADJUSTMENT T–SCORE

Predictions: Multiple Linear Regression (Causal Models)

Research Question: How, and how well do manifest aggression, psychopathic aggression, school failure and empathy predict social maladjustment?

Null Hypothesis: Manifest aggression, psychopathic aggression, school failure and empathy are not good predictors of social maladjustment?

Research Hypothesis: Manifest aggression, psychopathic aggression, school failure and empathy are not good predictors of social maladjustment?

Technical Interpretation

Descriptive statistics: The sample size for the regression was reasonable at about 228 respondents. The means for the interval-ratio variables indicate that on average social maladjustment, manifest aggression and aggressive psychopathy scales had scores of 60.47, 46.69, and 47.58 respectively.

Model summary: The model summary or [adjusted] R2 value indicates that about 60% of the variance in the DV, social maladjustment is predicted by manifest aggression, psychopathic aggression, school failure and empathy. About 40% remains unexplained.

Global F-test, ANOVA: The Global F-test is statistically significant (F = 84.896, $p <$ 0.001) clearly indicating that there are decent predictors in the model. Manifest aggression, psychopath aggression, school failure and empathy are 'decent' or 'good' predictors of social maladjustment score. We can continue our analyses.

Regression coefficients table unstandardized βs: The exact nature of all predictors is discussed here. For every one unit increase in manifest aggression, social maladjustment score increases by 1.053. Those that do not fail school, their social maladjustment score decreases by 3.386. Respondents that are empathetic, score 2.181 less than their non-empathetic counterparts on social maladjustment score. For every one unit increase in aggressive psychopath scale, social maladjustment score increases by 0.241. In summary, as expected, as aggression increases, social maladjustment scores also increase. However, those that are empathetic and did not fail school show improvement in their social maladjustment scores. These findings are not surprising and were expected.

Standardized Betas: The order of 'best predictor' results in the following outcome: Manifest aggression, followed by school failure, followed by aggressive psychopath, and lastly empathy are the best predictors, in that order, of social maladjustment score. Interestingly, school failure beat aggressive psychopath scale by a small amount. As the correlations and or associations were mapped out, our order of best predictors is similarly following similar patterns. However, surprisingly, empathy was deemed not significant in the final outcome.

Sample statistic: t-value and significance

The sample statistic t = 14.968, < 0.001 ($p <$ 0.05), t = −2.237, 0.026 ($p <$ 0.05), and t= −2.022, 0.044 ($p <$ 0.05) for manifest aggression, school failure and aggressive psychopath respectively were statistically significant predictors of social maladjustment. Empathy, surprisingly, was not significant (t = −1.223, 0.223, $p >$ 0.05). But was one of the weakest correlation values at the onset of the testing of this model. There is partial rejection of the null.

Predictions: Multiple linear regression (causal models)

Collinearity diagnostics: The Tolerance and Variance Inflation Factor (VIF) both indicate no instances of multicollinearity. All IVs or predictors are bringing their own unique variation to the model and there are no sharing violations amongst the IVs. All tolerance values are above 0.40 and all VIF values are less than 2.50. Similarly, the

Pearson Product Moment Correlation matched this outcome. As there no value greater than 0.80 between the predictors or IVs.

Outliers and graphs: The Casewise Diagnostics indicated that there are four cases in total with extreme scores. Mahal's Distance maximum value is 18.981. Because there are four predictors in the model, the critical value is 18.47. This only exceeds our value slightly by 0.51. Thus, there may not be multivariate outliers in the sample. Cook's distance indicates that outliers do not have an undue influence on the model as the value is less 1; it is 0.091. Anything larger than 0.50 is considered problematic. The histogram and normal p-plot show normality and an almost straight-line relationship. The final scatterplot of residuals also shows some clustering, thus there is a slight heteroscedastic stance. The constant spread of data points does is somewhat compromised. Thus, to ensure assumptions are met, perhaps removal of the four cases may improve the data and model fit.

Substantive Interpretation

Overall, the multiple linear regression model tested for four predictors against the DV, social maladjustment. This regression demonstrated that aggression, manifested or psychopathic, both are significant predictors and key variables in predicting social maladjustment. Likewise, being a school failure also predicts social maladjustment significantly well. Empathy surprisingly was not a significant predictor. Aggressive individuals need to increasingly mindful of social adjustments in the social world. Increasingly aggressive individuals are more likely to be socially maladjusted. The sample size at 228 may be less representative of the prison population. Therefore, take some precautions in generalizing the data.

Final Thoughts

Bivariate and multivariate regression are both straight line relationship models based on prediction and go one step further in answering many questions we have about predictability regarding the world around us. It provides us with data outcomes regarding causal linkages. Simple and complex predictions are tested by regression analyses. The goal is to explain and predict by exploring the underlying essence of the relationship at hand by not overfitting the model or making it too complicated. Model specification is key to ensuring that. How they do this varies. Bivariate models focus on one predictor at a time and multivariate models, slightly complex, test more than one predictor at a time. Both measure direct effects to the DV. Goodness of fit and model fit of these models are also tested. Regression analysis, compared to comparison of means test, provides valuable insights about the data and provides enhanced attention to detail about each specific predictor variable and as well provide information on the explanatory power of the model. While regression proves to be a powerful statistical analysis, it comes with its limitations. First, bivariate, and multiple linear regression assesses direct effects on a particular outcome. Also, it only analyzes a single DV or criterion variable. Additionally, this statistical technique is not assumption free and to produce reliable and valid outcomes, certain assumptions like linearity, independence of errors, homoscedasticity, multicollinearity, outliers, and normality must be met for data to be generalizable. Finally, regression falls to the basic principles of the Central Limit Theorem and therefore require adequate sample sizes to run and produce reliable and valid data with minimal Type 1 and Type 2 errors.

Hierarchical or Incremental or Block Modelling Multiple Regression

Another method of doing regression is referred to as Hierarchical regression. This popular type of regression analysis also comes under the purview of 'advanced' statistical techniques and has multiple names like hierarchical regression or incremental or block modelling regression analysis and provides slightly more advanced information, compared to Multiple linear regression, from a data analysis perspective. This regression follows similar principles to Multiple linear regression, as it predicts and explains, but does so by assessing and evaluating 'sets of predictors' is a systematic manner by examining the incremental contribution of sets of predictors to establish how much predictor variables contribute to the explained variance in the DV.

	Linear Multiple Regression	*Hierarchical Regression*
Variables and order of entry	Simultaneous entry of predictors and DV	Hierarchical or sequential entry of predictors and DV
Objectives	To assess simultaneous predictive power of the model	To assess 'sets' of predictors and see how the model improves
Interpretation	Done simultaneously, while controlling for the effects of all other predictors with no additions	Focuses on the unique contribution of 'sets' of predictors in an additive model
Models	Overall fit of one model	Overall fit, as well as improvement of different model with different predictors
Conceptualization	Must be conceptually sound and theory based	Must be conceptually sound and theory based

In Hierarchical regression, the researcher works in a specific conceptual order or hierarchy and builds models that are theoretically sound and plausible. The purpose is to assess the nature of the relationship as well as the variability in the dependent variable by examining specific 'sets' or 'block' or 'increments' of predictor variables in the model and their direct effects, while controlling for a certain set of variables. Variables are entered into model in a stepwise manner. The end goal of this type of regression modelling is to do the following:

- To identify the 'best' predictor in the model; the interest here is on the predictors by examining the standardized beta coefficients.
- Investigate 'sets' of predictors rather than individual predictors; this allows us to assess variability and identify the best set of predictors for a particular criterion/outcome variable and allow us to control for certain variables that may matter in the analysis.
- Variables are added in 'increments' based on proper logic and conceptualization of variables. Similar variables are put into their own 'blocks' or 'increments' to see how the explanatory power of the model changes or not.

Essentially, this is an ordinary least square (OLS) regression-based analysis that takes the hierarchical structure of the data into account. Again, the main end goal of this type of regression modelling is to evaluate relationships amongst 'sets' of predictor variables and the dependent variable, while controlling for a specific set of variables. The order of entry of each block is at the discretion of the researcher, but most often is theoretically inclined. The variables selected show make conceptual sense and have a theoretical underpinning. This type of regression modelling is based on logic and reason, with a solid social explanation. The first or second block or even final block entered this regression model may include control variables. Basically, those variables that we want to hold constant and ensure the variability of any control variable is removed from the analysis of predictor variables and the dependent variable. For example, in this type of regression we can examine what the effect of family support and social support is on intimate partner violence, after controlling for race and income. This is a two-block model that uses family and social support variables in Block 1 and race and income in Block 2, as potential controls. This multi-level modelling is unique, compared to simple linear regression that simultaneously place all predictors together. A hierarchical regression can have as many blocks as there are independent variables, i.e., the analyst can specify a hypothesis that specifies an exact order of entry for variables. A more common hierarchical regression specifies two blocks of variables: a set of control variables entered in the first block and a set of predictor variables entered in the second block. Control variables are often demographics which are thought to make a difference in scores on the dependent variable. Predictors are the variables in whose effect our research question is really interested, but whose effect we want to separate out from the control variables. In this regression, researchers rely heavily on theoretical conceptualization of their models. The hierarchy or order in which predictors are entered is of utmost importance. Each predictor can be part of one model or block and not multiple blocks. Thus, they cannot overlap. Basically, this type of regression wants to assess how additional sets of predictors improve the model or not? Unique contributions of each set are analyzed and discussed. The key changes in the model are observed and written about.

If we want to assess three models/blocks and make predictions on Criminal Behavior, this is the regression equation:

Block 1: Demographics or control variables, like gender and race
Block 2: Family support, like live with parents and parental neglect
Block 3: Social support, good friends and good teachers
Demographics Block 1: $\hat{y} = a + (b_1x_1 + b_2x_2) +$ error (2 predictors)
Family Support Block 2: $\hat{y} = a + (b_1x_1 + b_2x_2) + (b_3x_3 + b_4x_4) +$ error (2 more predictors)
Social Support Block 3: $\hat{y} = a + (b_1x_1 + b_2x_2) + (b_3x_3 + b_4x_4) + (b_5x_5 + b_6x_6)$ error (2 more predictors)

One popular question surrounding this type of regression modelling is how do we decide which order to enter the predictors or IVs in? Most often, the first model, is where incorporation of demographic variables, such as gender, race, years of education, marital status, etc. is done. This first block of predictors most likely influences the dependent variable. The next block or model we include our key predictors and in the final block we introduce the predictor variables that are highly correlated with the dependent variable. Here, the end goal is to see how well each additional block explains the outcome or DV above and beyond the other variables included in the models.

SPSS shows the statistical results (Model summary, ANOVA, Coefficients, etc.) as each block of variables is entered into the analysis. In addition, SPSS tests the key statistic used in evaluating the hierarchical hypothesis: change in R^2 or coefficient of determination for each additional block of variables. If the null hypothesis is rejected, then our interpretation indicates that the variables in block 2 had a relationship to the dependent variable, after controlling for the relationship of the block 1 variables to the dependent variable. In this regression model, the contribution of [adjusted] R^2 is assessed for each block or model to see if addition of predictor variables to the model improves the model or not and if the addition of variables is statistically significant; it also provides you with the 'change in R^2' to see how certain sets of predictors change the value from model to model. The change in explained variance by the addition of new variables assesses the DV, based on each block and should result in a change that is statistically significant. The unique contribution of each set of predictors to the DV is evaluated. If the explained variance increases with each additional block, then it works in favor of our model. If not, then predictors may have to reevaluated or reassessed and the model run again. Not only should the explained variance increase with the addition of blocks but should also be statistically significant.

As all inference-based tests, there are core assumptions that must be tested and met. The assumptions of this regression analyses are like that of Multiple Linear Regression. It works under similar assumptions.

Activity Alert

Discuss some of the characteristics of Hierarchical Regression?
How does it differ from Multiple Linear Regression?

Assumptions	Explanation
1. Data or Level of measurement of variables	Regression analysis assumes that the independent and dependent variables are interval-ratio continuous or its equivalent (i.e., scaled, or dummy coded; 1 is the reference group)
2. Sampling method and sample size	Random sampling and large sample
3. Independence of observations	Data observations must be independent of each other; no influence by other observations
4. Shape of the distribution	Normally distributed of bivariate relationship such that bivariate normality is met for each set of IV and DV
5. Homoscedasticity	Constant spread of variance (equal variances for residuals or error terms)
6. Absence of multicollinearity	The IVs are not highly correlated with each other; each IV brings its own unique variance; Each correlation is less than 0.80. The IVs are not highly correlated with each other; each IV brings its own unique variance; Each correlation is less than 0.80. Variance Inflation Factor > 2.50; or Tolerance Values < 0.40
7. Linearity	The relationship between the IVs and DV is linear
8. Confidence level	The amount of risk you are willing to take: 95%, 0.05 alpha
9. Sample statistic	t-statistic or t-value; rejects the H_o or fails to reject it to see overall significance of the IVs alongside DV
10. Outliers	Minimize outliers or extreme scores using Casewise Diagnostics, Mahalanobis Distance and Cooks Distance

The research question and hypotheses for are written as the following:

Research Question: After controlling for sex and age, can marital status and education predict number of children?

Null and Research hypotheses

Null hypothesis: After controlling for sex and age, marital status and education cannot predict respondents' number of children. There are no statistically significant predictors or IVs for the DV.

Research hypothesis: After controlling for sex and age, marital status and education can predict respondents' number of children. There are no statistically significant predictors or IVs for the DV. There are statistically significant predictors or IVs for the DV.

The sample statistic, the t-value, like Multiple Linear regression, determines whether there are significant predictors in each block. However, the final block is most important in determining the outcome. Most often, in analyzing hierarchical regression we report the final model or block, as that considers any control effects from the previous model. Examination of all models is essential to see if the patterns in the statistical reporting changes or not.

Key Statistics to Report

1   Descriptive statistics, sample size, averages, means, and standard deviation are reported so that trends and patterns of the variables can be understood; you would assess all interval-ratio or scaled variables

2   Adjusted R^2 (explained variance) or the coefficient of determination explains how much variance in the DV is being explained by certain predictors out of 100%; the change in Adjusted R2 and whether the change is statistically significant

3   ANOVA Global F-test (decent predictors in the model) and must be statistically significant; this indicates that there are decent predictors in the model or the way the model has been conceptualized is correct; if this value is not significant then the model needs to be reconceptualized

4   Regression Coefficients for each model or 'sets' of predictors; analyze final model

    a   Unstandardized β (beta) tells the story of each predictor with respect to the DV; it tells us for every unit increase/decrease in x, y increases by a certain amount; if the value is positive, it is an increase; if the value is negative it is decrease; the unstandardized β is in units of the DV always

    b   Standardized β (beta) taking the absolute value (regardless of signage), tells the *order* of best predictors as it relates to the DV. They are on the same scale and thus can be compared. Unstandardized β's cannot do this

    c   Sample statistic: t-value is the sample statistic that is the measure that assesses statistical significance of each regression coefficient as it relates to the DV. This statistic tests the null hypothesis and rejects or fails to reject the null hypothesis. Using a 95% confidence level, if the $p < 0.05$ then statistical significance is achieved and the null hypothesis is rejected; if the $p > 0.05$, then statistical significance is not achieved, and one fails to reject the null hypothesis

    d   Collinearity diagnostics: assess multicollinearity among IVs (predictors) in a regression model. Multicollinearity occurs when two or more predictors have a sharing violation, and no unique variance is present. The correlation amongst predictors is strong to include in the model. Correlation (> 0.80), Variance Inflation Factor (> 2.50) and Tolerance (< 0.40) indicate multicollinearity.

5   Outliers: Casewise Diagnostics, Mahalanobis Distance (Mahal's D) and Cook's Distance

6   Extreme scores, high or low, in the data: Mahal's Distance critical value must not be exceeded; Cook's Distance any value greater than 1.00 indicates outliers make an undue influence
7   Review Histogram and Normal p-p plot for examining normality and straight-line relationships for the predictors and DV and residuals

Hierarchical Regression in SPSS

1   Click on Analyze > Regression > Linear
2   Move the variables of interest into the Dependent (DV) and Independent Variable Boxes as 'sets of predictors' by Blocks (Block 1 of 1, Block 2 of 2, etc....)
3   Click on Statistics > Check off Estimates, Model Fit, Descriptives, R^2 change and Collinearity Diagnostics; Check-off Casewise Diagnostics and Click on Continue
4   Click on Plots > Y: Add ZPRED and X: Add ZRESID > Next check off Histogram and Normal Probability Plot > Click on Continue
5   Click on Save > Check off Mahalanobis Distance and Cook's Distance
6   OK to generate a Hierarchical Regression Model

Compared to Multiple linear regression, Hierarchical regression is much more sophisticated in dealing with predictors and large, complex data. The way they both model causal linkages amongst variables differ significantly. In Multiple linear regression, the goal is to simultaneously build a relationship or model without a specific order in mind for predictors. In Hierarchical regression, the blocks or increments of each set of predictors is the essence in how it relates to the DV. There is a definitive hierarchy or order of each block or increment entry with logic and reason of a specific theory or social explanation in mind. In It provides much more meaningful statistics, especially with respect to how each set of predictors behaves in each block. This information becomes quite useful in analyzing a specific outcome or measure. By showing the delta or change in blocks as sets of predictors are added to the model provides clarity on the explanatory power of each block and its fit. If there is a choice given between multiple linear regression or hierarchical regression, the latter is much more informative as it considers the hierarchical impact of control variables as well.

No test is without limitations. One of the biggest limitations is that this test requires a large sample size and has many assumptions to fulfill for reliable and valid data. With real data, sometimes meeting all assumptions deems to be problematic and violations may result in inaccuracies of outcomes and findings. Each assumption needs to carefully be evaluated to ensure that the model and model fit is accurate. Additionally, how blocks or sets of predictors are entered into the model may bias our results. Thus, one must be methodologically careful in choosing the order of entry for each block. Do not misinterpret the findings or focus solely on statistical significance. That is only one aspect of the results. Examine the model and each aspect of the data to have an accurate picture of what is happening in the models. All key statistics need to be carefully interpreted to understand results, model fit and any biases. This type of analysis is sensitive to extreme scores or outliers and therefore some carefulness must be taken into consideration when making decisions about how to proceed. The best outcomes are when outliers are removed from the analysis. Reliability of findings is heavily linked to all assumptions being met. In the case where an assumption has been violated, findings should be generalized carefully from a sample to a population and any violation should be discussed as a limitation.

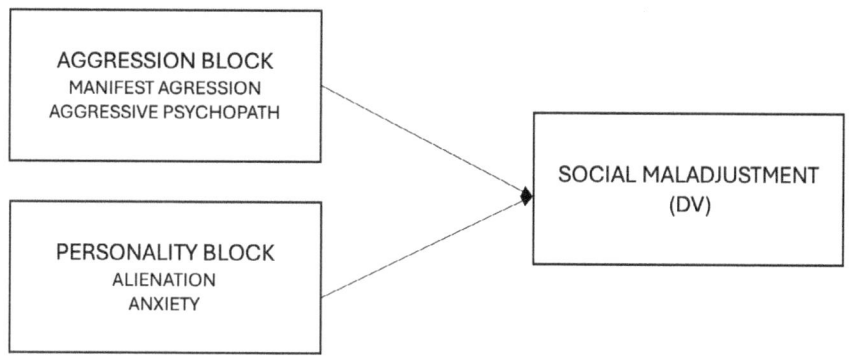

Figure 11.4a Causal Model of Hierarchical Regression of Blocks of Aggression, Personality Indicators as Predictors of Social Maladjustment

Analyze	Graphs	Utilities	Extensions	Window	Help

Power Analysis	>	Q Search application			
Meta Analysis	>				
Reports	>		Values	Missing	Columns
Descriptive Statistics	>	{1, Yes}...	9	8	≣ Ri
Bayesian Statistics	>	{1, Yes}...	9	7	≣ Ri
Tables	>	{0, 0:undo...	9	7	≣ Ri
Compare Means and Proportions	>	{1, Yes}...	9	7	≣ Ri
General Linear Model	>	{97, More ...	99	9	≣ Ri
Generalized Linear Models	>	{0, Felony ...	999	9	≣ Ri
Mixed Models	>	{0, Felony ...	999	9	≣ Ri
Correlate	>	{0. Felony ...	999	9	≣ Ri
Regression	>	⬚ Automatic Linear Modeling...			ⵑ
Loglinear	>	⬚ Linear...			ⵑ
Neural Networks	>	Linear OLS Alternatives	>		ⵑ
Classify	>	⬚ Curve Estimation...			ⵑ
Dimension Reduction	>	⬚ Partial Least Squares...			ⵑ
Scale	>				
Nonparametric Tests	>	⬚ Binary Logistic...			ⵑ
Forecasting	>	⬚ Multinomial Logistic...			ⵑ
Survival	>	⬚ Ordinal...			ⵑ
Multiple Response	>	⬚ Probit...			ⵑ
⬚ Missing Value Analysis...					ⵑ
Multiple Imputation	>	⬚ Nonlinear...			ⵑ
Complex Samples	>	⬚ Weight Estimation...			ⵑ
⬚ Simulation...		⬚ 2-Stage Least Squares...			ⵑ
Quality Control	>	⬚ Quantile...			ⵑ
Spatial and Temporal Modeling...	>	⬚ Optimal Scaling (CATREG)...			ⵑ
Direct Marketing	>	⬚ Kernel Ridge...			ⵑ
)RE					
–SCORE		{99. Missin...	99	5	≣ Ri

Figure 11.4b SPSS Command for Hierarchical Regression

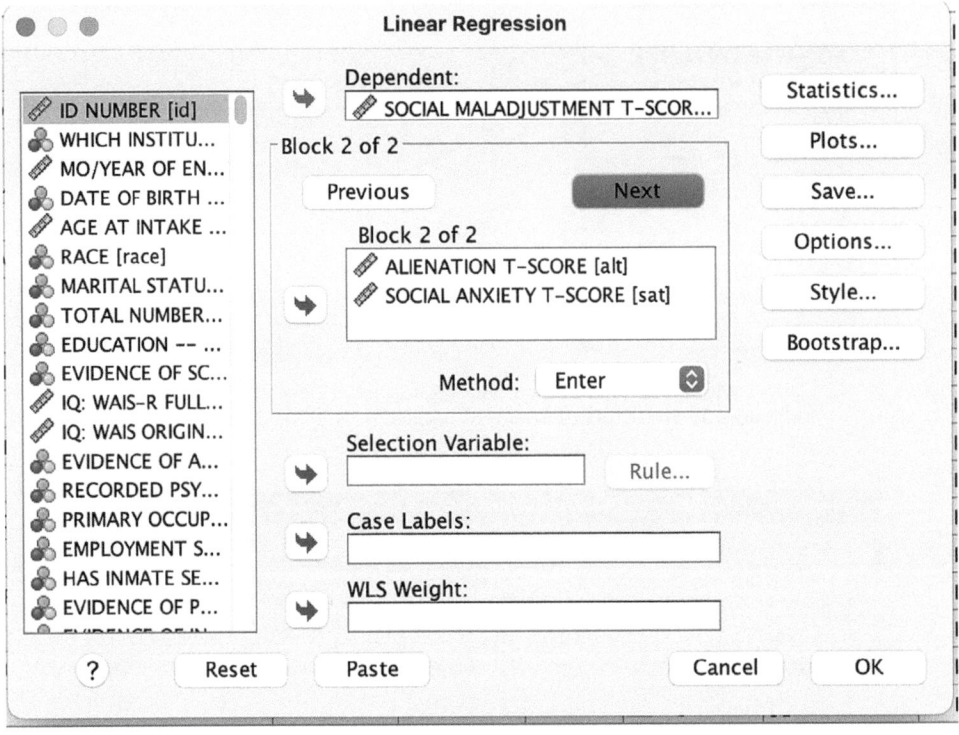

Figure 11.4c Move the variables of interest into the Dependent (DV) and Independent Variable Boxes as 'sets of predictors' by Blocks (Block 1 of 1, Block 2 of 2, etc....)

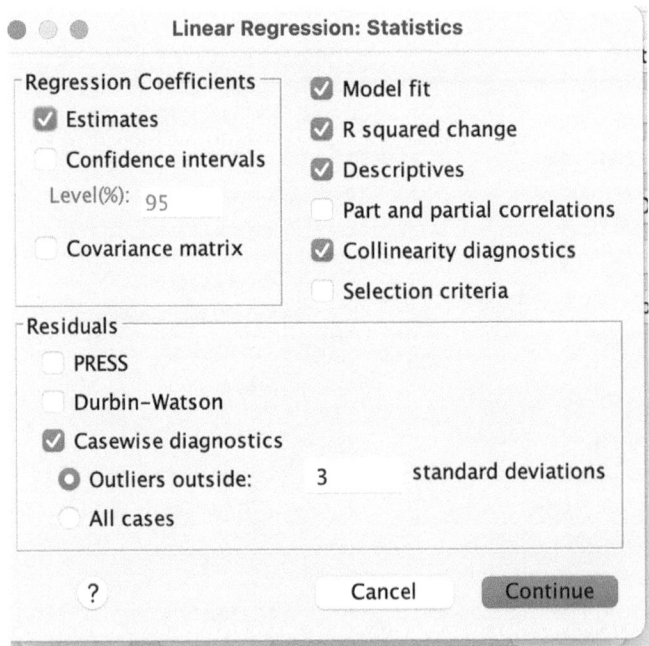

Figure 11.4d Click on Statistics and Check off Estimates, Model Fit, Descriptives, R2 change and Collinearity Diagnostics; Check-off Casewise Diagnostics and Click on Continue

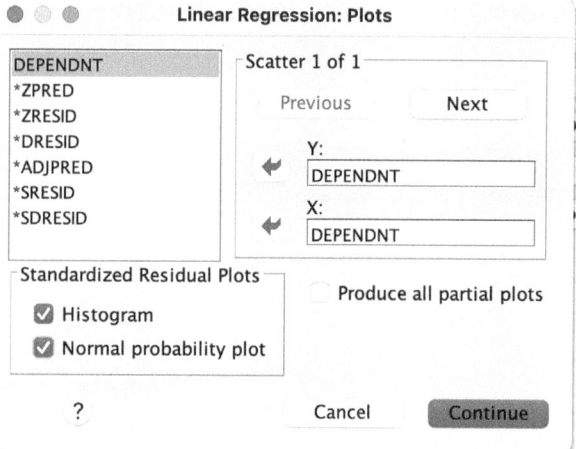

Figure 11.4e Click on Plots, Y: Add ZPRED and X: Add ZRESID and Check-off Histogram and Normal Probability Plot and Click on Continue

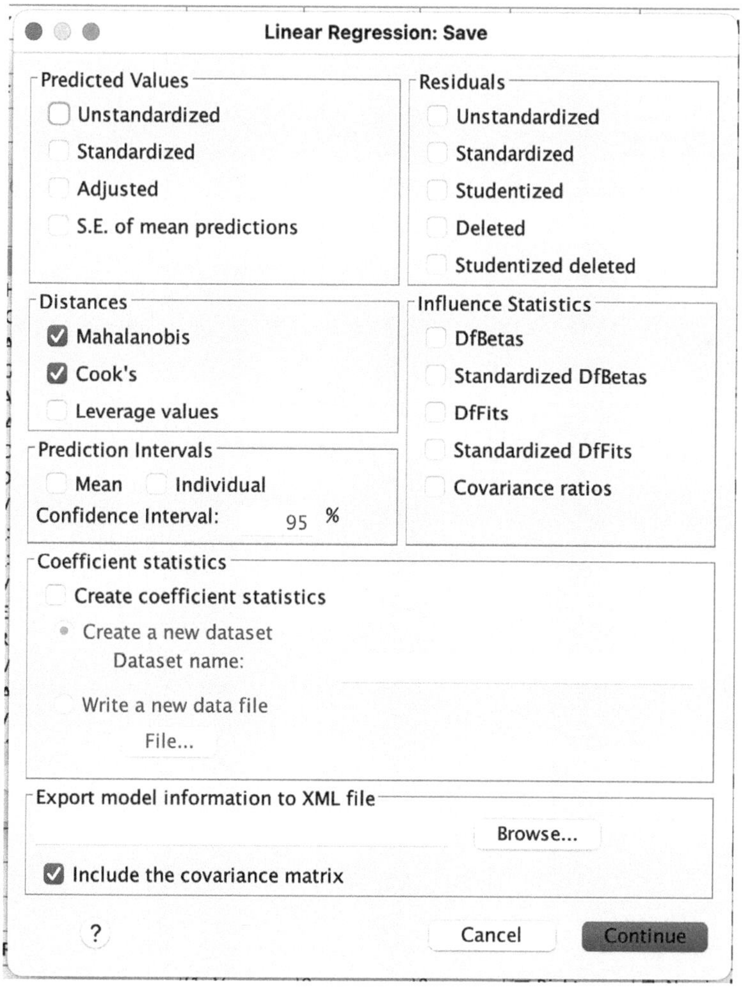

Linear Regression: Save

Predicted Values
- Unstandardized
- Standardized
- Adjusted
- S.E. of mean predictions

Residuals
- Unstandardized
- Standardized
- Studentized
- Deleted
- Studentized deleted

Distances
- ☑ Mahalanobis
- ☑ Cook's
- Leverage values

Influence Statistics
- DfBetas
- Standardized DfBetas
- DfFits
- Standardized DfFits
- Covariance ratios

Prediction Intervals
- Mean Individual
Confidence Interval: 95 %

Coefficient statistics
- Create coefficient statistics
  - ● Create a new dataset
    Dataset name:
  - Write a new data file
    File...

Export model information to XML file
Browse...
- ☑ Include the covariance matrix

?   Cancel   Continue

Figure 11.4f Click on Save and Check-off Mahalanobis Distance and Cook's Distance

Regression

Descriptive Statistics

	Mean	Std. Deviation	N
SOCIAL MALADJUSTMENT T–SCORE	60.68	15.934	242
MANIFEST AGGRESSION T–SCORE	46.81	10.346	242
T SCORE FOR AGGRESSIVE PSYCHOPATH SCALE	47.70	6.709	242
ALIENATION T–SCORE	58.26	11.095	242
SOCIAL ANXIETY T–SCORE	41.37	10.877	242

Figure SPSS Output #11.2 Hierarchical or Incremental Regression

Model Summary[c]

					Change Statistics				
Model	R	R Square	Adjusted R Square	Std. Error of the Estimate	R Square Change	F Change	df1	df2	Sig. F Change
1	.768[a]	.590	.586	10.246	.590	171.900	2	239	<.001
2	.821[b]	.674	.668	9.176	.084	30.512	2	237	<.001

a. Predictors: (Constant), T SCORE FOR AGGRESSIVE PSYCHOPATH SCALE, MANIFEST AGGRESSION T–SCORE

b. Predictors: (Constant), T SCORE FOR AGGRESSIVE PSYCHOPATH SCALE, MANIFEST AGGRESSION T–SCORE, SOCIAL ANXIETY T–SCORE, ALIENATION T–SCORE

c. Dependent Variable: SOCIAL MALADJUSTMENT T–SCORE

ANOVA[a]

Model		Sum of Squares	df	Mean Square	F	Sig.
1	Regression	36093.618	2	18046.809	171.900	<.001[b]
	Residual	25091.241	239	104.984		
	Total	61184.860	241			
2	Regression	41231.286	4	10307.821	122.432	<.001[c]
	Residual	19953.574	237	84.192		
	Total	61184.860	241			

a. Dependent Variable: SOCIAL MALADJUSTMENT T–SCORE

b. Predictors: (Constant), T SCORE FOR AGGRESSIVE PSYCHOPATH SCALE, MANIFEST AGGRESSION T–SCORE

c. Predictors: (Constant), T SCORE FOR AGGRESSIVE PSYCHOPATH SCALE, MANIFEST AGGRESSION T–SCORE, SOCIAL ANXIETY T–SCORE, ALIENATION T–SCORE

Coefficients[a]

Model		Unstandardized Coefficients		Standardized Coefficients			Collinearity Statistics	
		B	Std. Error	Beta	t	Sig.	Tolerance	VIF
1	(Constant)	-7.367	4.927		-1.495	.136		
	MANIFEST AGGRESSION T–SCORE	1.071	.069	.695	15.602	<.001	.864	1.158
	T SCORE FOR AGGRESSIVE PSYCHOPATH SCALE	.376	.106	.158	3.548	<.001	.864	1.158
2	(Constant)	-24.248	5.406		-4.486	<.001		
	MANIFEST AGGRESSION T–SCORE	.566	.092	.368	6.163	<.001	.387	2.586
	T SCORE FOR AGGRESSIVE PSYCHOPATH SCALE	.319	.099	.134	3.238	.001	.800	1.250
	ALIENATION T–SCORE	.612	.079	.426	7.790	<.001	.459	2.178
	SOCIAL ANXIETY T–SCORE	.182	.063	.124	2.912	.004	.756	1.323

a. Dependent Variable: SOCIAL MALADJUSTMENT T–SCORE

Casewise Diagnostics[a]

Case Number	Std. Residual	SOCIAL MALADJUSTMENT T-SCORE	Predicted Value	Residual
19	4.334	81	41.23	39.771
42	3.964	81	44.63	36.368
60	4.339	81	41.18	39.815
118	4.439	81	40.27	40.730
278	4.772	81	37.22	43.785

a. Dependent Variable: SOCIAL MALADJUSTMENT T-SCORE

Residuals Statistics[a]

	Minimum	Maximum	Mean	Std. Deviation	N
Predicted Value	34.58	90.37	60.68	13.080	242
Std. Predicted Value	-1.996	2.270	.000	1.000	242
Standard Error of Predicted Value	.673	2.672	1.277	.329	242
Adjusted Predicted Value	34.39	90.38	60.67	13.091	242
Residual	-27.127	43.785	.000	9.099	242
Std. Residual	-2.956	4.772	.000	.992	242
Stud. Residual	-2.968	4.826	.001	1.003	242
Deleted Residual	-27.348	44.787	.011	9.301	242
Stud. Deleted Residual	-3.019	5.072	.004	1.022	242
Mahal. Distance	.301	19.446	3.983	2.645	242
Cook's Distance	.000	.144	.004	.015	242
Centered Leverage Value	.001	.081	.017	.011	242

a. Dependent Variable: SOCIAL MALADJUSTMENT T-SCORE

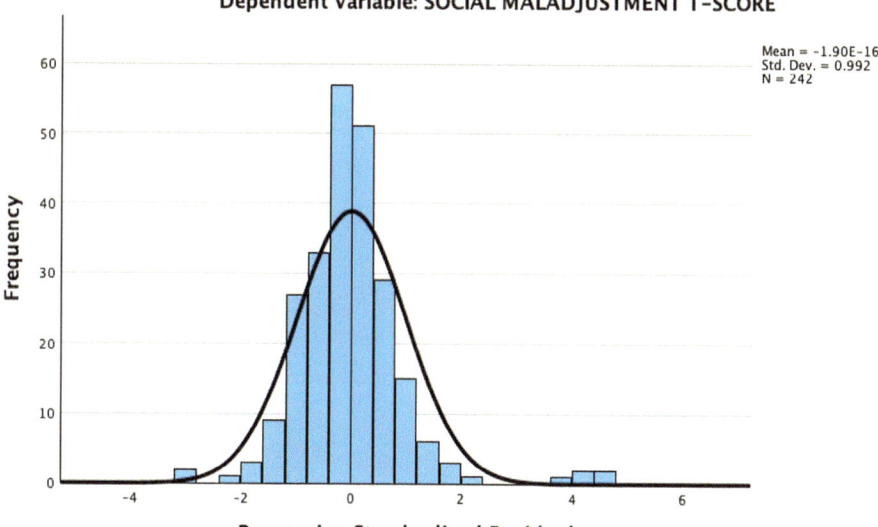

Histogram
Dependent Variable: SOCIAL MALADJUSTMENT T-SCORE

Mean = -1.90E-16
Std. Dev. = 0.992
N = 242

Frequency

Regression Standardized Residual

Normal P–P Plot of Regression Standardized Residual
Dependent Variable: SOCIAL MALADJUSTMENT T–SCORE

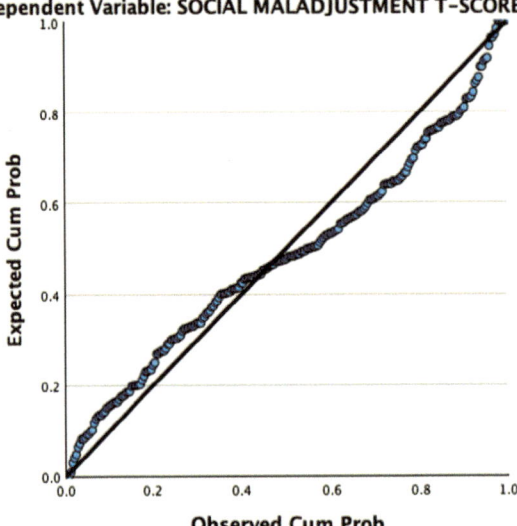

Scatterplot
Dependent Variable: SOCIAL MALADJUSTMENT T–SCORE

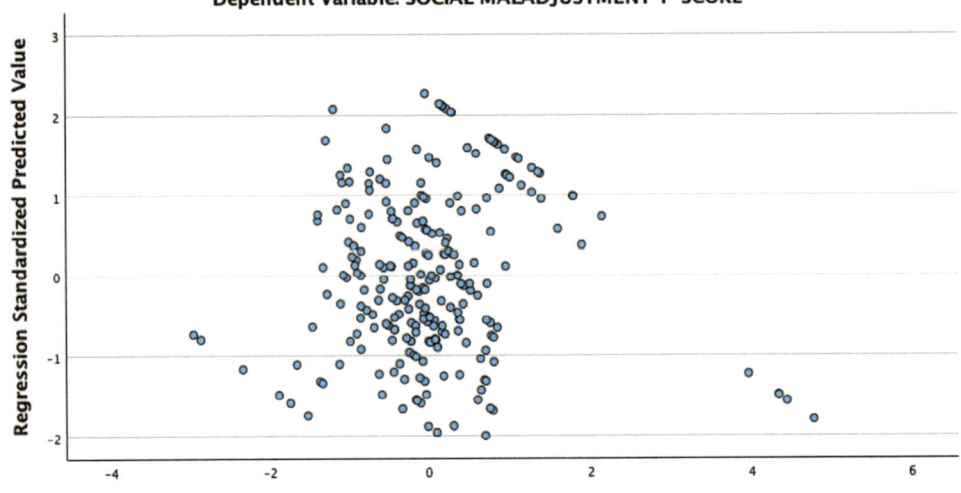

Predictions: Hierarchical Regression (Causal Models)

Research Question: How, and how well does the aggression block and personality traits block predict social maladjustment?

Null Hypothesis: After controlling for manifest and aggressive psychopathy, alienation and social anxiety cannot predict social maladjustment. There are no statistically significant predictors or IVs for the DV and no change is R^2 values.

Research Hypothesis: After controlling for manifest and aggressive psychopathy, alienation and social anxiety cannot predict social maladjustment. There are no statistically significant predictors or IVs for the DV and no change in R^2 value.

Technical Interpretation

Descriptive statistics: The sample size for the incremental regression was 242, The means for the interval-ratio continuous variables indicate that on average social maladjustment, manifest aggression, aggressive psychopathy, alienation, and social anxiety score scales had scores of 60.68, 46.81, 47.70, 58.26 and 41.37 respectively. This regression is a two-block model, first controlling for the effects of aggression and then analyzing personality traits, like alienation and anxiety.

Model summary: The model summary or [adjusted] R^2 value indicates that about 58.6% of the variance in the DV, social maladjustment is being accounted for by Block 1: Aggression; 66.8% of the variance in the DV, social maladjustment is being accounted for by Block 2: Personality Traits. The addition of variables or blocks to the model is statistically significant. Thus, the R^2 change is noteworthy (F = 171.900, $p <$ 0.05; F = 30.512, $p <$ 0.05). The significant R2 change or difference in the adjusted R^2 is about 8.2% change. The addition of Block 2 personality predictors makes a difference. While a 30–40% difference in unexplained variances between the blocks goes unaccounted for, the amount both these blocks account for is reasonably well with respect to social maladjustment.

Global F-test, ANOVA: The Global F-test for both increments or blocks are statistically significant (F = 171.900, $p <$ 0.001; F = 122.432, $p <$ 0.001) clearly indicating that both blocks comprise of decent predictors in the model. We can continue our analyses.

Regression coefficients table unstandardized βs: The exact nature of all predictors in each block are discussed here. For every one unit increase in manifest aggression, social maladjustment score increases by 0.566. For every one unit increase in aggressive psychopath scale, social maladjustment score increases by 0.319. For every one unit increase in alienation score, social maladjustment increases by 0.612; and finally for every one unit increase in social anxiety score, social maladjustment score increases by 0.182. Thus, aggression and personality traits increase social maladjustment scores slightly. These findings are not surprising and were expected.

Standardized betas: The order of 'best predictor' results in the following outcome: Manifest aggression, followed by alienation score, followed by aggressive psychopath, and lastly, social anxiety score. Interestingly, manifest aggression and alienation were the top two predictors of social maladjustment. It is noteworthy to see how some predictors on the aggression block and personality block both have causal linkages to social maladjustment.

Predictions: Hierarchical regression (causal models)

Sample statistic: t-value and significance: The sample statistic for t = 6.163, p < 0.001 (manifest aggression), t = 3.238, $p <$ 0.001 (aggressive psychopath), t = 7.790, p < 0.001 (alienation score), and t = 2.912, $p <$ 0.001 (social anxiety score) for both increments or blocks. All predictors are statistically significant. Thus, there is full rejection of the null hypothesis. Even after controlling for the aggression block, personality traits block remained statistically significant.

Collinearity diagnostics: The Tolerance and Variance Inflation Factor (VIF) both indicate that there for the most part there no major instances of multicollinearity. All IVs or predictors are bringing their own unique variation to the model and there are no sharing violations amongst the IVs. However, alienation and aggression collinearity diagnostics were a little high. But because one is measuring aggression and the other is measuring anxiety, so they may not be overlapping so much as to impact the model. The impact is very slight.

Outliers and graphs: The Casewise diagnostics indicated that there are five cases in total with extreme scores. Mahal's Distance maximum value is 19.446. Because there are 4 predictors in the model, the critical value is 18.47. This only exceeds our value slightly by 0.976. Thus, there may not be multivariate outliers in the sample. Cook's distance indicates that outliers do not have an undue influence on the model as the value is less 1; it is 0.081. Anything larger than 0.50 is considered problematic. The histogram and normal p-plot show normality and an almost straight-line relationship. The final scatterplot of residuals also somewhat of a constant spread of variance, with little clustering. The constant spread of data points does very slightly compromise. Thus, to ensure assumptions are met, perhaps removal of the four cases may improve the data and model fit.

Substantive Interpretation

Overall, the hierarchical regression model tested for two blocks or increments against the DV, social maladjustment. This regression demonstrated that while controlling for the aggression block, personality traits also were significant key predictors of social maladjustment. Increasingly aggressive individuals are more likely to be socially maladjusted. The sample size at 242 may be less representative of the prison population. Therefore, take some precautions in generalizing the data.

Binary and Multiple Logistic Regression

So far, we have discussed regression analyses that allows us to make predictions are solely based on linear straight-line relationships or the ordinary least squares (OLS) method with both IV and DV being interval-ratio continuous measures or its equivalent. Both Multiple linear regression and Hierarchical regression tests follow a similar trajectory, especially with regards to their assumptions and data requirements and are not very flexible in their approach. Both these methods are excellent and hold their own ground, but their lack of flexibility in terms of assumptions and levels of measurement makes both these techniques difficult to work with, especially with real-life data. Real-life data comes with real-life problems and not always are their ideal data sets with continuous level variables. In most cases, data are binary, categorical, and dichotomous. The main objective of Logistic regression is to classify individuals into groups of belonging or not belonging or passing or failing or criminal or not using non-continuous data.

Logistic regression is different and unique in so many ways, comparatively speaking. It predicts a binary DV outcome, such as yes or no. Using a dichotomous DV, makes the nature of prediction unusual and non-linear. In this type of regression, the relationship between the IV(s) and the probability of the event occurring (the log-odds or logit) is the focal point. Here, the normal curve is not the standard distribution followed, but the Sigmoid Curve or Logistic function is most appropriate. This function transforms the log-odds into a range between 0 and 1, thus making predictive probabilities a possibility. This type of regression predicts the DV by analyzing its relationship with an IV (Binary logistic regression); it can be bivariate or multivariate in nature and have multiple IVs (Multiple logistic regression). Many independent and dependent variables we want to understand do not occupy interval-ratio continuous measurement. Instead, they are in a binary, dichotomous and dummy coded state of 0 and 1, in which 1 is the reference group. This indeed is very different than previous OLS statistical methods. In doing survey research, most often, the researcher has variables that are dichotomous and binary in nature. They are categorical in nature and often identified as the dependent variable in nature.

For example, a survey question like, have you ever been arrested by police in the year 2023? [0] no [1] yes, is a great example of a dichotomous binary dummy coded DV. This question can be tested against IVs, like race, gender, years of education, or immigrant status. In this type of regression, your IV can assume categorical properties, discrete or interval-ratio continuous. This is a great benefit. If you are a running a multivariate model, then your IVs can be a mixed level of measurement model or heterogeneous with respect to levels of measurement. The versatility of Logistic regression is impressive. Its popular due to its flexibility towards the nature and type of data that is used. Not always are normally distributed data available and not always are the highest levels of measurement are accessible. When there is a disconnect of linearity and level of measurement, the Logistic regression proves to be fruitful. Modelling binary outcomes is the strength of Logistic regression. In fact, Logistic regression is best used when there is:

a   Dependent variable is dichotomous categorical
b   When assumptions of linearity, normality and homoscedasticity cannot be met
c   When you wish to know the predictive probability of a particular event occurring
d   Basically, when you have bad data, Logistic regression is the method of choice. Also, when you want to make predictive probabilities.
e   To rank the relative importance of predictor variables in explaining the outcome
f   To calculate the ODDS ratio that measures the importance of a predictor variable on that outcome.

Logistic regression's main objective is to calculate the probabilities of the outcomes (i.e., like criminal or not; pass or fail; successful or not). The value predicted in Logistic regression is a probability ranging from 0 to 1, absence of an attribute or presence of attribute. It is most often identified as an assumption free test, unlike multiple linear regression models. Non-linear data works best with Logistic regression. All values in Logistic regression range from positive 0 to 1 and there are no negative probabilities possible. Thus, Logistic regression is based on the *Logit model* which is the natural log of the Odds. It becomes increasingly large in absolutely value and allows us to fix the problem of estimated probability exceeding 0 or 1. This also corrects for non-linearity in relation between IV and DV. There are not many conditions when running a logistic regression, however a few noteworthy suggestions are to have no cell counts with 0, or greater than 20% of cells with less than five participants; no outliers and best outcomes if N is greater than or equal to 100.

The functionality of Logistic regression is countless. This test can do the following:

1   Assess how well your predictor(s) predicts or explains the categorical DV
2   Allows us to predict which two categories a person is likely to belong to given certain information
3   Gives an indication of adequacy to the model
4   Provides an indication of the relative importance of each predictor

5   Provides summary of accuracy of the classification of cases
6   Provides the explained variance of the model

Logistic regression's main goal is to predict group membership by calculating the probability of success. While this is an assumption free test, some core assumptions are to have a large enough sample size so accurate predictions can be made. Intercorrelations amongst your predictors are frowned upon, just as seen in any regression technique discussed. There should always be a lack of multicollinearity among the predictors. Finally, any outliers should be removed, especially if the Goodness of fit tests do not work out.

Odds, Odds Ratio and Probability

Odds ratios (OR) are derived from the coefficients and represent the change in the odds of the event occurring for a one-unit change in the predictor variable. Any OR greater than 1 suggests an increased likelihood, while an OR less than 1 suggests a decrease in likelihood.

> **Probability**: The ratio of the number of occurrences to the total number of possibilities.
> Probability = Odds / 1 + Odds
> **Odds**: Probability that an event will occur divided by the probability that it will not occur can range from 0 to infinity.
> Odds = Exp $^{(A + B(x))}$
> **Odds ratio**: The change in odds of being in one of the categories of outcome when
the value of a predictor increases by one unit (Tabachnick and Fidell, 2013).

Activity Alert

Discuss some of the characteristics of Logistic Regression?
What is the key dependent variable difference in Logistic Regression?
Why is Logistic Regression considered to be flexible?

Key Statistics to Report

1   Classification: tells us the overall percentage of correctly classified cases
2   Compare Classification tables in Block 0 and 1 to see improvement in predicting model based on adding IVs.
3   Chi-Square in Omnibus tests of model coefficients
4   Goodness of fit test: Compares model with all predictors to a model with no predictors; similar to F-test in multiple linear regression; determines whether model accounts for a significant amount of variance in DV
5   −2 log-likelihood based on the summation of predicted probabilities and actual probabilities looking at unexplained variance
6   Pseudo R^2 similar to [Adjusted] R^2 in linear regression; gives us information regarding how much variation our predictors account for; could be used as an effect size for the model. Two main types: Cox and Snell R2 and Nagelkerke R2
7   Wald Statistic tells us the statistical significance of individual predictors in the logistic regression model.
8   Odds Ratio (OR) quantifies the odds that an outcome will occur given the presence or absence of a particular association.

Binary Logistic Regression in SPSS

1 Click on Analyze > Regression > Binary Logistic Regression
2 Move the DV to the Dependent Box
3 Add all your categorical dummy coded Predictors to the Covariate(s) box
4 In Save > Check off Probabilities and Group Membership
5 In Options > Check off Classification Plots, Hosmer-Lemeshow Goodness of fit, Casewise listing of residuals, and CI for Exp

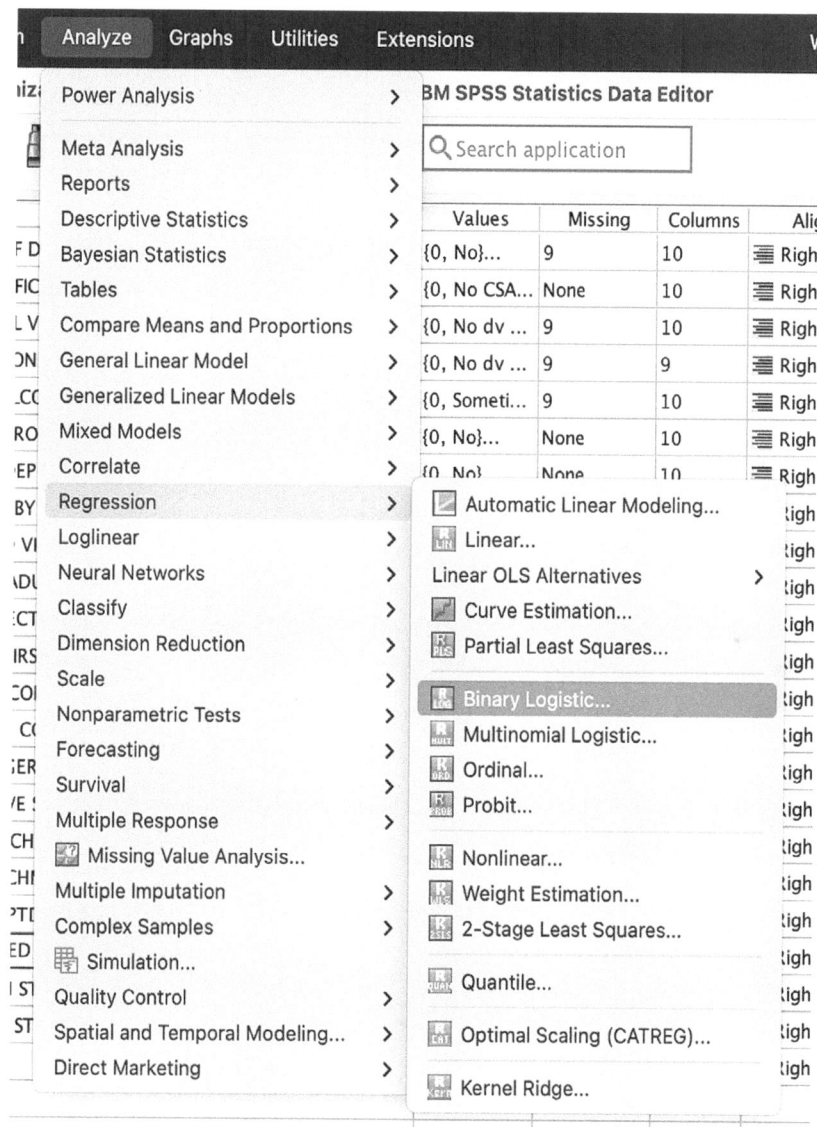

Figure 11.5a SPSS Command for Binary Logistic Regression and move the DV to the Dependent Box; Add all your categorical dummy coded Predictors to the Covariate(s) box

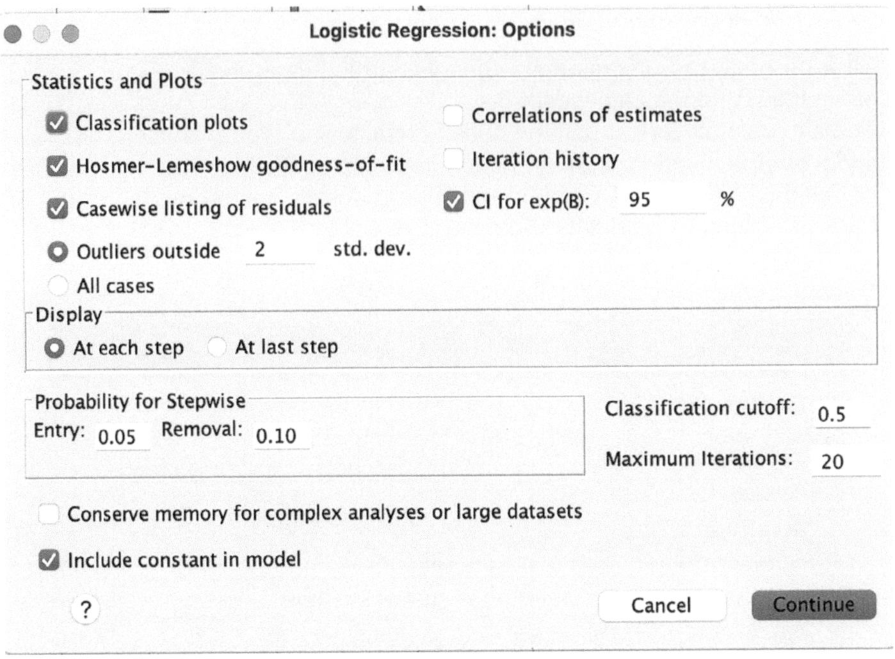

Figure 11.5b In Options, Check-off Classification Plots, Hosmer-Lemeshow Goodness of fit, Casewise listing of residuals, and CI for Exp

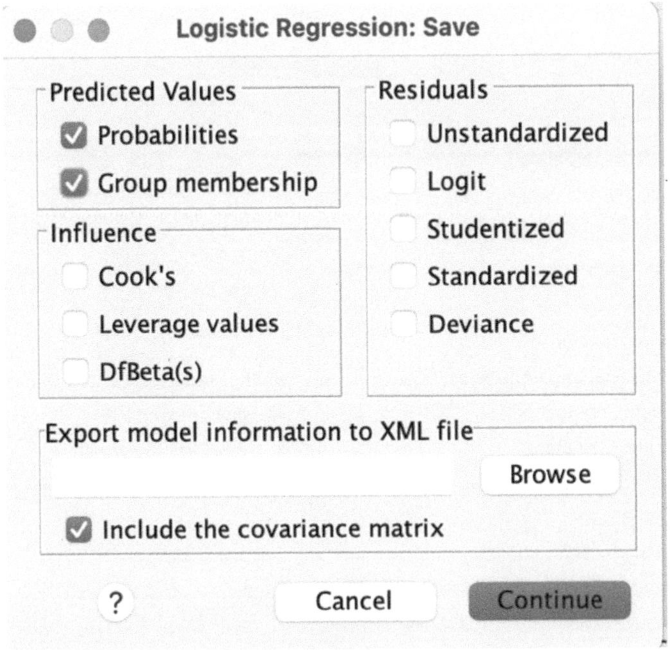

Figure 11.5c In Save, Check-off Probabilities and Group Membership

Logistic Regression

Case Processing Summary

Unweighted Cases[a]		N	Percent
Selected Cases	Included in Analysis	168	96.6
	Missing Cases	6	3.4
	Total	174	100.0
Unselected Cases		0	.0
Total		174	100.0

a. If weight is in effect, see classification table for the
total number of cases.

Dependent Variable Encoding

Original Value	Internal Value
No	0
Yes	1

Categorical Variables Codings

		Frequency	Parameter coding (1)
EVER ALCOHOL DEPENDENCY (1=yes)	No	116	.000
	Yes	52	1.000
EVER PROSTITUTED (1=yes)	No	131	.000
	Yes	37	1.000

Block 0: Beginning Block

Classification Table[a,b]

			Predicted		
			88. PARENTS DRUNK/HIGH PROB TAKING CARE		Percentage Correct
	Observed		No	Yes	
Step 0	88. PARENTS DRUNK/HIGH PROB TAKING CARE	No	147	0	100.0
		Yes	21	0	.0
	Overall Percentage				87.5

a. Constant is included in the model.

b. The cut value is .500

Figure SPSS Output #11.3 Binary Logistic Regression

Variables in the Equation

		B	S.E.	Wald	df	Sig.	Exp(B)
Step 0	Constant	-1.946	.233	69.578	1	<.001	.143

Variables not in the Equation

			Score	df	Sig.
Step 0	Variables	EVER PROSTITUTED (1=yes)(1)	9.155	1	.002
		EVER ALCOHOL DEPENDENCY (1=yes)(1)	1.592	1	.207
	Overall Statistics		9.381	2	.009

Block 1: Method = Enter

Omnibus Tests of Model Coefficients

		Chi-square	df	Sig.
Step 1	Step	8.085	2	.018
	Block	8.085	2	.018
	Model	8.085	2	.018

Model Summary

Step	-2 Log likelihood	Cox & Snell R Square	Nagelkerke R Square
1	118.510[a]	.047	.089

a. Estimation terminated at iteration number 5 because parameter estimates changed by less than .001.

Hosmer and Lemeshow Test

Step	Chi-square	df	Sig.
1	.027	2	.987

Contingency Table for Hosmer and Lemeshow Test

		88. PARENTS DRUNK/HIGH PROB TAKING CARE = No		88. PARENTS DRUNK/HIGH PROB TAKING CARE = Yes		Total
		Observed	Expected	Observed	Expected	
Step 1	1	91	91.159	8	7.841	99
	2	29	28.841	3	3.159	32
	3	13	12.841	4	4.159	17
	4	14	14.159	6	5.841	20

Classification Table[a]

			Predicted		
			88. PARENTS DRUNK/HIGH PROB TAKING CARE		Percentage
Observed			No	Yes	Correct
Step 1	88. PARENTS DRUNK/HIGH PROB TAKING CARE	No	147	0	100.0
		Yes	21	0	.0
	Overall Percentage				87.5

a. The cut value is .500

Variables in the Equation

		B	S.E.	Wald	df	Sig.	Exp(B)	95% C.I.for EXP(B) Lower	Upper
Step 1[a]	EVER PROSTITUTED (1=yes)(1)	1.326	.508	6.813	1	.009	3.766	1.391	10.192
	EVER ALCOHOL DEPENDENCY (1=yes)(1)	.242	.511	.224	1	.636	1.274	.468	3.468
	Constant	-2.453	.347	50.057	1	<.001	.086		

a. Variable(s) entered on step 1: EVER PROSTITUTED (1=yes), EVER ALCOHOL DEPENDENCY (1=yes).

Binary Predictions: Multiple Logistic Regression

Research Question: What predictive factors, alcohol dependency and prostitution, predict the likelihood that respondents would report that their parents are drunk or high and have a problem with taking care of them?

Null Hypothesis: Alcohol dependency and prostitution are not predictors of parents drunk or high and have a problem with taking care of them.

Research Hypothesis: Alcohol dependency and prostitution are predictors of parents drunk or high and have a problem with taking care of them.

Technical interpretation

Descriptive statistics: The sample size for the logistic regression is 174 with 96.6% cases included in the analysis. There are total of 6 or 3.4% missing cases. This model contained two IVs or predictors, alcohol dependency and prostitution and one DV, parents drunk or high and have a problem of taking care of them. This is a multiple logistic regression with coding of the DV or criterion variable in those that do not have parents drunk or high and have issues taking care is 0, and those that have parents drunk or high and have issues taking care is 1.

Model summary of block 0 or the classification table: The baseline model is assessed here. This is a model that does not consider predictor variables. The predictions here are based on the baseline model and depend on which category most often occurred. In this model, 'no' was more, compared to yes for parents being drunk or high and having a problem with taking care. The overall percentage of prediction is 87.5% correct. This is much greater than a coin toss.

Variables in the equation and variables not in the equation: The variables in the equation table simply expose the coefficient for the constant in Step 0. This table while not so informative is suggestive that the model with only the constant indicates a statistically significant predictor of the outcome ($p < 0.05$) and the accuracy is about 88% of the time. Variables not in the equation simply report back which predictors were not in the equation.

Omnibus tests of model coefficients: This test assesses 'model fit'. With the addition of variables, the model shows a significant improvement in fit as compared to Block 0;

the overall models show good fit when we add out explanatory or predictor variables to the model (X 2 = 8.085, $p < 0.05$).

The model summary and Hosmer Lemeshow Test: The Loglikelihood is 118.510. The Nagelkerke R2 suggests that 8.9% of change in the DV, can be accounted by alcohol dependency and prostitution. The Hosmer and Lemeshow Test indicates that the model successfully fits the data as its outcome is not significant (X 2 = 0.027, $p > 0.05$). The differences between the observed and expected values are similar or approximately equal. Again, the classification table suggests that with predictors, the overall accuracy rate was about 87.5% and considered good.

Variables in the equation: This table provides the regression coefficient and the Wald statistic and Odds Ratio for each variable. There is a high significant effect for ever prostituted (Wald = 6.813, $p < 0.05$), but a non-significant effect of ever alcohol dependent (Wald = 0.224, $p > 0.05$). The Exp(B) column or the Odds Ratio that if you ever prostituted than you are 3.766 times, compare to those that do not to have high or drunk parent taking care of them. However, only 1.274 times for those that are alcohol dependent to have high or drunk parents taking care of them.

Substantive Interpretation

Overall, this model suggests that prostitution as a predictive variable is a highly significant variable with the DV parents high or drunk and problem taking care. Alcohol dependency was not. While the model fit was good, the overall variation of the model was less than 10% with respect to explanatory power. Other predictors should be tested for, like race, age of respondent, ran away from home, victim of domestic violence, etc. and see what the predictive probabilities are. The sample size was not very large, and some groups were underrepresented. This may have skewed our findings slightly. Any generalizations should be treated cautiously.

Summary Linear Multiple, Hierarchical and Logistic Regression Types

	Linear multiple regression	Hierarchical regression	Logistic regression
Variables and order of entry	Interval-ratio continuous IV(s) and DV or its equivalent or simultaneous entry of predictors and DV	Interval-ratio continuous IV(s) and DV or its equivalent or hierarchical or sequential entry of predictors and DV	Categorical, discrete, interval-ratio continuous IV(s) and dichotomous, binary or categorical DV
Objectives	To assess simultaneous predictive power of the model	To assess 'sets' of predictors and see how the model improves	To assess 'probability' of an event occurring and see how the log-odds (logit) operates
Interpretation	Done simultaneously, while controlling for the effects of all other predictors with no additions	Focuses on the unique contribution of 'sets' of predictors in an additive model	Focuses on the Odds ratio and to calculate the probabilities of the outcomes
Models	Overall fit of one model	Overall fit, as well as improvement of different model with different predictors	Overall fit and how well model explains the relationship between the IV and DV via predictive accuracy
Conceptualization	Must be conceptually sound and theory based	Must be conceptually sound and theory based	Must be event or outcome based

Final Thoughts

This chapter discussed various types of regression models and addressed the similarities and differences between each test. Each regression analysis provides a story-telling of causal linkages and establishes cause and effect to a certain degree. Linear multiple regression, Hierarchical regression and Logistic regression are the main regression techniques discussed at length. It is important to understand the similarities and differences each test exhibits. While the end goal of each regression test discussed is similar, the temperaments of each regression analysis is unique and flourishes differently. Each regression analysis comes with its own unique model and answers questions regarding forecasting, estimation, optimization, and predictions differently. Compared to all the other inferential statistical tests that we have covered, regression analyses trumps all of them, especially in how it establishes cause and effect. Its versatility and power of analysis towards understanding the IV(s) and DV of each model is applauded and appreciated. One of the biggest drawbacks of regression analyses is its list of never-ending assumption checks. The assumption checks are vast and must be fulfilled to achieve a reliable and valid regression outcome. Logistic regression tries to overcome this requirement but has its limitations in capturing complex relationships. Linear regression models and hierarchical models are the ultimate tests but follow a stringent pathway to establish causality. Each regression method comes with its own merits and flaws, and it becomes most important to understand when to use which method. While there are many regression options in the social sciences, the best regression outcomes result when models are fully understood to meet the statistical expectations. Failure to not follow assumptions, results in increased error and a flawed analytical design in which conclusions are meaningless and unreliable. There always needs to be great thought put into the conceptualization and re-conceptualization of the regression model to achieve the best outcomes in statistics. Ultimately, regression analysis is the end goal for any social scientist. This is the statistical test, given that all assumptions are securely met, we try to get at it with our IV (s) and DV. The power and efficiency of this test is like no other discussed in this book. While there are many statistical choices we can make depending on the levels of measurement of our variables, regression analysis is a great test to run as its attention to detail given to each IV and DV is phenomenal. The numerical storytelling done by regression tests surpasses all of them.

Keywords and Definitions

Linear multiple regression/Ordinary least squares method (OLS) (Bivariate or Multiple)	A very sophisticated test which puts in predictor variables simultaneously into the model and tells the researcher the change in Y (i.e., the dependent variable) that is associated with a change in X (i.e., independent variable). It is designed to help predict the most likely value for the other variable based on available information.
Hierarchical or Incremental regression analysis or Block modelling	This takes in a conceptually specified model in steps/blocks by grouping variables together. Again, this is like linear multiple regression analysis; the key difference here is that variables are NOT put in simultaneously, but rather in a very theoretically conceptualized manner.

Logistic regression	This is different and unique in so many ways, comparatively speaking. It predicts a binary DV outcome, such as yes or no. This type of regression predicts the DV by analyzing its relationship with an IV (Binary Logistic Regression); it is bivariate or multiple IVs (Multiple Logistic Regression). Many independent and dependent variables we want to understand do not occupy interval-ratio continuous status. Instead, they are in a binary, dichotomous, dummy coded state of 0 and 1.
Adjusted R^2	Explained variance, or the coefficient of determination, explains how much variance in the DV is being explained by certain predictors out of 100%; the change in Adjusted R^2 and whether the change is statistically significant
ANOVA Global F-test	(decent predictors in the model) and must be statistically significant; this indicates that there are decent predictors in the model or the way the model has been conceptualized is correct; if this value is not significant then the model needs to be reconceptualized
Unstandardized β (beta)	This tells the story of each predictor with respect to the DV; it tells us for every unit increase/decrease in x, y increases by a certain amount; if the value is positive, it is an increase; if the value is negative it is decrease; the unstandardized β is in units of the DV always.
Standardized beta	Often denoted as β, this takes the absolute value (regardless of signage) and tells the order of best predictors as it relates to the DV. They are on the same scale and thus can be compared. Unstandardized β's cannot do this.
Goodness of fit test	In logistic regression models, this compares a model with all predictors to a model with no predictors. It is similar to the F-test in multiple linear regression and determines whether model accounts for a significant amount of variance in DV.
Pseudo R^2	Similar to [Adjusted] R^2 in linear regression, this gives us information regarding how much variation our predictors account for. It can be used as an effect size for the model. There are two main types: Cox and Snell R^2 and Nagelkerke R^2.

Test Your Knowledge

1 What are the types of regression the social sciences are blessed with?

  a Hierarchical
  b Bivariate
  c Logistic
  d Multivariate
  e All of the above

2 The ____ determines the amount of variation explained in the DV by various predictors in a model.

    a   Wald Statistic
    b   Odds Ratio
    c   Coefficient of determination or Adjusted R^2
    d   Slope of the line
    e   Standardized beta

3 Logistic regression is an assumption free test. True or False?

    a   True
    b   False

4 Which regression is heavily dependent on interval-ratio continuous data for both the IVs or predictors and DV?

    a   Logistic regression
    b   Multiple linear regression
    c   Hierarchical regression
    d   a and b
    e   b and c

5 What are the assumptions of multiple linear regression and hierarchical regression analysis?

    a   Normality, linearity, lack of multicollinearity, random sampling, homoscedasticity
    b   Normality, multicollinearity, non-random sampling, heteroskedasticity
    c   Having theoretically specified models establishing causal linkages
    d   Variance inflation factor and tolerance values
    e   These regressions are assumption free

6 What is the difference between unstandardized βs and standardized βs?

    a   Unstandardized βs tell the story for every unit increase or decrease
    b   Standardized βs allow us to compare and provide order of best predictors
    c   Unstandardized βs reject the null hypothesis
    d   Standardized betas are a model fit test
    e   a and b only

7 A research hypothesis for regression analyses is best written like this:

    a   To assess average differences in predictor variables, gender, race and education on income
    b   To test a relationship in predictor variables, gender, race, and education on income
    c   To assess, how and how well do predictor variables, gender, race, and education significantly predict income?
    d   To test how multiple predictors co-vary?
    e   To assess statistical independency of predictor variables?

8 $\hat{y} = a + a + \beta_1 x_1 + \beta_2 x_2 \beta_3 x_3 + \beta_n x_n +$ error prediction equation belongs to what type of regression model?

   a Logistic regression
   b Multiple regression
   c Hierarchical regression
   d Goodness of fit regression
   e None of the above

9 Linear regression works on the principles of the central limit theorem?

   a True
   b False

10 Regression analyses are the most sophisticated inferential tests working with the highest levels of measurement.

   a True
   b False

Notes

1 Logistic regression is unique as it predicts a binary dependent variable (nominal), compared to a continuous dependent variable; moreover, it requires no assumptions. This type of regression is based on the Logistic function, rather than a straight-line relationship. A Hierarchical regression model is based on 'sets of predictors' and does not include all predictor variables simultaneously in one block. Instead, this type of regression modelling is based on model comparison of various 'blocks' or 'sets of predictors' in the model, while controlling for other variables. Their needs to be conceptual specification of each block and what predictors are included. The assumptions for this type of regression are like Multiple Linear regression. Again, they differ in terms of how variables are entered into the model.
2 Logistic regression is a probabilistic regression model that is a type of analysis that does not require any core assumptions to be met as the DV is a dichotomous categorical variable with values of 0 and 1. It does not follow normal distribution properties or characteristics.
3 Mahalanobis Distance for assessment of outliers is based on exceeding the critical value based on the number of IVs or predictors. For example, if there are 2, 3, 4, 5, 6, 7 predictors in a model, the corresponding critical values are: 13.82, 16.27, 18.47, 20.52, 22.46, 24.32, respectively (Field, 2024).

Appendix: Revised Tables for Peer Reviewed Articles, Theses, or Dissertations

Refashioning SPSS Outputs to Revised Tables for Publication

The biggest challenge of doing statistics is to convert SPSS outputs to refined, redone, and readable tables. These tables should illustrate the most relevant information that tells the story of the data. Here are some examples of how the information from outputs can be easily transferred. Obviously, you decide as the researcher or writer what relevant information is needed in each table. I have provided some common ways data are presented in theses, dissertations, papers, conference presentations or articles. Here are some examples of statistical tests that we frequently engage in. These tables, again, provide some guidance towards final table reporting in statistics. You can always add to the general gist of the table, depending on what the research objective (s) is, what is being presented and discussed. This, however, provides a platform of getting started. Often getting started is the most difficult task.

Table A.1 Univariate Trends and Patterns for a Nominal Categorical Variable – "Subject Threw Something at Partner that Hurt" (N = 174)

Threw something at partner that hurt	Frequency	Valid percentage
[0] No	99	57.2%
[1] Yes	74	42.8%
TOTAL	173	100.0

Notes: Mode: 0 Missing Cases: 1

Table A.2 Zero-Order Crosstabulation of How Physical Abuse by either Parent varies with Ever a Victim of Domestic Violence, while Controlling for Subject Ever a Perpetrator of Violence (N = 173)

Victim of domestic violence		Physical abuse by either parent	
		No abuse by parent	Yes, by at least one parent
	No	14	39
		50%	26.9%
	Yes	14	106
		50%	73.1%
Total		100%	100%

Notes: Zero-Order Crosstabulation Pearson Chi-Square = 5.895, :015, $p < 0.05$
Measure of Association: .185, 0.015, $p < 0.05$
Elaborated Crosstabulation Pearson Chi-Square$_{no}$ = 2.828, .093, $p > 0.05$
Elaborated Crosstabulation Pearson Chi-Square$_{yes}$ = 1.316, .251, $p > 0.05$

Table A.3 An Exploration of "Average Differences" using an Independent Samples t-test of Sentence Length for Current Offence and Whether or not Inmate Served Prior Prison Sentence (N = 346)

	Sentence	Length (Months)	
	N	Mean	Standard deviation
Inmate served prior prison sentence			
[1] Yes	159	111.95	103.844
[2] No	187	83.65	155.293

Notes: Levene's Test for Equality of Variances: F = .073, *p* > 0.05, Equal variances assumed
t-value: t = 0.026, *p* < 0.05
Mean Difference: 28.303 months
Cohen's d: 0.211

Table A.4 A One-way ANOVA of Average of Total Number of Cigarettes Smoked in Past Seven Days and Mother's Approval of Such Behavior (N = 788)

	# Cigarettes			
	N	Mean	SD	Minimum/maximum
Mother's approval				
She approves	30	65.83	45.730	6–175
She doesn't care	110	59.01	41.703	1–190
She doesn't like it	275	39.30	35.873	0–210
She doesn't know that I smoke	373	15.26	25.446	0–252
Total	788	Grand mean 31.68		

Notes: Levene's Test for Equality of Variances: F = 35.318, *p* < 0.05, Not equal variances assumed
F-value: F_{welch} = 64.374, *p* < 0.001**

Table A.5 Two-way ANOVA of Average of Total Number of Cigarettes Smoked in Past Seven Days and Main and Interaction Effects of Mother and Father's Approval (N = 715)

Main and interaction effects	F	Significance	Partial eta^2
Mother's approval	6.496	< 0.001**	0.027
Father's approval	4.820	0.002*	0.020
Mother*father approval	2.031	0.041	0.023

Notes: Adjusted R^2: .265
Levene's Test for Equality of Variances: F = 8.802, *p* < 0.001, Not equal variances assumed
p < 0.05
**p* < 0.001

Table A.6 ANCOVA on Average Number of Children with Main Effect, Political Views, after Controlling for Age of Respondent as a Covariate (N = 1397)

Main & covariate effects	F	Significance	Partial eta^2
Age (covariate)	213.510	< 0.001**	0.133
Political views	4.820	0.394	0.004

Notes: Homogeneity of slopes: F = 1.265, *p* > 0.05 (Assumption met)
Adjusted R^2: 13.7%
Levene's Test for Equality of Variances: F = 0.290, *p* > 0.05, Not equal variances assumed
p < 0.05
**p* < 0.001

Table A.7 MANOVA Assessing Significant Mean Differences in the Combined DVs, Months Since Hire and Previous Experience, for Factors, Minority Classification and Job Category (N = 948)

Main and interaction effects	F	Significance	Partial eta^2
DV – Months Since Hire (months)			
Multivariate Tests: Wilks' Lambda			
Job category	30.527	< 0.001**	
Minority classification	0.995	0.370	
Job category*Minority classification	5.8323	< 0.001**	
Between subjects effects			
Job category	2.674	0.070	0.011
Minority classification	1.908	0.168	0.004
Job category*Minority classification	3.780	0.024*	0.016
Adjusted R^2			0.8%
DV- Previous Experience (months)			
Multivariate Tests: Wilks' Lambda			
Job category	30.527	< 0.001**	
Minority classification	0.995	0.370	
Job category*Minority classification	5.8323	< 0.001**	
Between subjects effects			
Job category	61.803	< 0.001**	0.209
Minority classification	0.085	0.770	0.000
Job category*Minority classification	7.935	< 0.001**	0.033
Adjusted R^2			24.9%

Notes: *p < 0.05
**p < 0.001
Box's M and Levene's Test: p > 0.05, Equal variances assumed

Table A.8 Pearson Product Moment Correlations (r) of Predictor Variables, Manifest Aggression, School Failure, Empathy and Aggressive Psychopath Scale on Social Maladjustment (N = 228)

	Manifest aggression (x_1)	School failure (x_2)	Empathy (x_3)	Aggressive psychopath (x_4)
Social maladjustment (y)	+.756**	–0.302**	–0.203*	+0.421**

Notes: *p < 0.05
**p < 0.001
No issues of multicollinearity evident, r< .80

Table A.9 Manifest Aggression, School Failure, Empathy, Psychopath Aggression as Predictors or Regressors of Social Maladjustment Tendencies (N = 228)

Predictor variables (IVs)	Unstandardized b	Standardized Beta	t-value
Manifest aggression	1.053	0.686	14.968**
School failure [1 = not failure]	−3.386	−0.105	−2.237*
Empathy	−2.181	−0.053	−1.223
Aggressive psychopath	0.241	0.102	2.022*
Adjusted R^2: 59.7%			

Notes: *$p < 0.05$
**$p < 0.001$

Table A.10 Hierarchical Regression of a Two Block Causal of Aggression, Personality Indicators as Predictors of Social Maladjustment (N = 242)

Predictor variables (IVs)	Unstandardized b	Standardized Beta	t-value
Aggression block			
Manifest aggression scale	0.566	0.368	6.163**
Aggressive psychopath scale	0.319	0.134	3.238*
Personality block			
Alienation score	0.612	0.426	7.790**
Social anxiety score	0.182	0.124	2.912*

Notes: ANOVA Global F-tests, $p < 0.05$*
Adjusted R^2 model 1: 58.6%; model 2: 66.8% (8.2% difference)

Table A.11 Multivariate Logistic Regression Model of Ever Prostituted and Ever Alcohol Dependent as Predicting the Likelihood of Parents Drunk or High and Have an Issue with Taking Care of Them (N = 174)

	B	Wald Statistic	Significance	EXP (B)
Predictor Variables				
Ever prostituted	1.326	6.813	0.009*	3.766
Ever alcohol dependency	0.242	0.224	0.636	1.274

Notes: *$p < .05$
Nagelkerke R^2: 8.9%

References

Bachman, R., Paternoster, R., & Wilson, T. (2022). *Statistics for Criminology and Criminal Justice* (5th edn). Thousand Oaks, CA: Sage.

Creswell, J.D., & Creswell J.W. (2017). *Research Design Qualitative, Quantitative and Mixed Method Approaches* (4th edn). Newbury Park: Sage.

Denzin, N.K., & Lincoln, Y.S. (2000). *Handbook of Qualitative Research*. Thousand Oaks, CA: Sage Publications.

Field, A. (2024). *Discovering Statistics Using IBM SPSS Statistics*. Thousand Oaks, CA: Sage.

Hinton, P.R., McMurray, I., Brownlow, C., & Terry, C. (2024). *SPSS Explained* (3rd edn). Routledge.

Mertler, C.A., & Reinhart, R.A.V. (2017). *Advanced and Multivariate Statistical Methods: Practical Application and Interpretation* (6th edn). Routledge.

Meyers, L.S., Gamst, G., & Guarino, A.J. (2016). *Applied Multivariate Research: Design and Interpretation*. Sage Publications.

Neuman, L.W. (2013). *Social Research Methods: Qualitative and Quantitative Approaches* (7th edn). Pearson Canada Publishing Company.

Stockemer, D. (2019). *Quantitative Methods for the Social Sciences: A Practical Introduction with Examples in SPSS and Stata*. Springer.

Tabachnick, B.G., & Fidell, L.S. (2013). *Using Multivariate Statistics* (6th edn). Boston: Pearson.

Index